D1544776

China and the Geopolitics of Rare Earths

# China and the Geopolitics
## of Rare Earths

**SOPHIA KALANTZAKOS**

# OXFORD
UNIVERSITY PRESS

Library of Congress Cataloging-in-Publication Data
Names: Kalantzakos, Sophia, author.
Title: China and the geopolitics of rare earths / Sophia Kalantzakos.
Description: New York, NY, United States of America : Oxford University Press, [2018] |
Includes bibliographical references and index.
Identifiers: LCCN 2017019913 (print) | LCCN 2017022864 (ebook) | ISBN 9780190670948 (Updf) |
ISBN 9780190670955 (Epub) | ISBN 9780190670931 (hardcover : acid-free paper)
Subjects: LCSH: Rare earth industry—Political aspects. | Geopolitics. | Rare earth metals—China. |
China—Foreign economic relations.
Classification: LCC HD9539.R32 (ebook) | LCC HD9539.R32 K35 2018 (print) |
DDC 333.8/54940951—dc23
LC record available at https://lccn.loc.gov/2017019913

*To George Karampatzos (1945–2011)*

# CONTENTS

# ACKNOWLEDGMENTS

I owe a special debt to Jonathan Spence and Paul Kennedy for sparking my interest in Chinese history and in questions of geopolitics. It would be hard to imagine more discerning and compelling guides to the complexities faced by anyone trying to work on these topics. A chance encounter with Athanasios G. Konstandopoulos first alerted me to the importance of rare earths in contemporary technological applications, and I have subsequently benefited from many enjoyable discussions. Asteris Huliaras has been a careful and helpful critic of this book in all its stages, and I wish to thank him not only for his unstinting encouragement and guidance, but also for his warm friendship. Stephen Holmes, Phillip Mitsis, Lara Nettelfield, Andreas A. Papandreou, and Dimitri Psaras all offered helpful criticism and advice, while Vasanth Mohan has been an invaluable guide in navigating the complexities of the rare-earth markets. Beth Daniel Lindsay of the NYU Abu Dhabi Library has been an invaluable resource, and Otto Kakhidze, my former student, generously assisted in the preparation of the figures and tables.

A fellowship at the Rachel Carson Center enabled me to do the final revisions of the manuscript in idyllic conditions, and I am deeply grateful to its directors, Christof Mausch and Helmut Trischler, as well as to my cohort of fellows, for making my time in Munich so stimulating and pleasant. Over the years NYU and NYU Abu Dhabi have provided both time and funds in support of my research, and I am happy to take this opportunity to thank them.

I have been very lucky to have Angela Chnapko as my editor. She has been exemplary in every respect and I owe a further debt of thanks to OUP's anonymous readers for exceptionally helpful and detailed criticism.

Finally, I would like to dedicate this book to the memory of George Karampatzos who relentlessly tried to steer me away from the world of politics and into academia. He succeeded, and I definitely owe him my thanks.

China and the Geopolitics of Rare Earths

# Introduction

*Rare Earths: A Crisis in the Making*

In 2010, seventeen elements from the periodic table known as "rare earths"[1] suddenly became notorious overnight. Until that moment, few people outside the mining and tech industries had heard of these particular materials. Indeed, they had been undervalued for decades, even though they had already grown indispensable in the production of renewable energy resources and green technology (such as wind turbines, solar panels, and efficiency lighting), in high-tech applications (such as computers, smart phones, and medical applications), and in the defense industry (such as in missile guidance systems, smart bombs, and submarines). Abruptly, rare earths were transformed from simple inputs in modern applications to materials of strategic[2] and economic importance, worthy of front-page news. What was the catalyst for this change in not only the perception of rare earths, but also in criticality?

To begin with, rare earths are subject to domination by China to an extent unparalleled in the previous history of strategic materials. Although other strategic resources, such as oil, have been highly geographically concentrated,[3] the fact that one state alone now holds a near monopoly[4] on the production of such crucial resources is unprecedented. In fact, there is no other example of one nation having such a stranglehold on the supply of a vital element as is the case with China and rare earths. According to the European Union's Critical Materials List, published in 2011,[5] based on their supply risk and economic importance, it appears that worldwide production of additional resources—identified as "critical"—comes from just a handful of countries and China is dominant there, as well.

Furthermore, in the case of oil, the technology for both its extraction and processing is widespread. In the case of rare earths, the converse holds true. At the moment, and for the foreseeable future, China not only monopolizes at least 93 percent of these materials,[6] in a sense becoming an OPEC of one, but it also controls their highly specialized metallurgy and the entire supply chain, from mine to market, for many crucial applications.[7] This essentially gives China a stranglehold on a significant range of essential inputs of the world's economy, especially with respect to renewable energy, high-tech, and military applications.

The main catalyst for rare earths' sudden rise to prominence and as front-page news, however, was a single international incident that turned the global focus toward China's ability to use its near monopoly to further leverage its strategic aims or, at the very least, to provide a forceful response in a crisis situation. In September 2010,[8] the People's Republic of China "unofficially" halted rare-earth exports to Japan because of a maritime incident[9] in the disputed waters near the Senkaku Islands (or the Diaoyu Islands according to China) in the East China Sea.[10] This sent a shockwave throughout the international community because of worries that China might extend the embargo beyond Japan.[11] It was at that point that the United States,[12] through then Secretary of State Hillary Clinton, first stepped in to address the potential supply disruption, the repercussions of China's rare-earth monopoly, and its use for geopolitical ends.[13]

In a comprehensive and targeted visit to US allies in Asia, and, in particular, during her 2010 meeting with Japan's Foreign Minister Seiji Maehara, Clinton responded to questions about rare earths, emphasizing their importance in industry and security. She underscored that it was imperative that other nations look for sources of rare earths outside China to cut their reliance on one source alone. Specifically, she stated that

> these are elements that are critical to the industrial production not only in Japan and the United States but in countries around the world . . . because

of the importance of these rare earth minerals, I think both the Minister and I are aware that our countries and others will have to look for additional sources of supply. That is in our interests commercially and strategically; it makes sense because these are rare, by definition, but they are present in other countries. And this served as a wakeup call that being so dependent on only one source, disruption could occur for natural disaster reasons or other kinds of events could intervene.[14]

The Japanese foreign minister, in turn, addressed the issue in the following careful language:

> With regard to the rare earth minerals, as Secretary Clinton stated earlier, even if this problem did not exist, to rely for 97 percent of these resources on China, as we look back, was certainly not appropriate and therefore we have to diversify the sources of rare earth minerals. And here again, Japan and the United States will closely cooperate with each other in order to engage in more diversified rare earth minerals diplomacy.[15]

In 2010, moreover, one other critical change in China's management of rare earths had preceded the diplomatic incident. China had already been slowly cutting its rare earths export quotas starting in 2006, but by the summer of 2010, the cuts had reached 40 percent[16] in comparison to 2009, and the prices of the elements began to skyrocket. This dramatic change in China's rare-earth management strategy prompted growing concern in industries that depended on the uninterrupted supply of these materials. Oddly enough, however, the issue only exploded internationally, drawing the attention of governments, the world community, the media, and international business, after China was perceived as being willing to use its near monopoly to leverage a territorial dispute. The incident gave the industrial nations pause and brought them to the realization that it would perhaps be prudent to diversify their rare-earth sources.[17]

China's growing global activism, its newfound economic strength and political influence, and its long-term strategic planning have raised concerns among industrial nations. They monitor China, engage with China, and trade with China, but they are not exactly sure what kind of superpower China will be and how its rise will affect the current status of power relations and global leadership paradigms. For years, China has built its international relations on the concept of maintaining good working relationships with other countries so that it can concentrate on its domestic transformation. The PRC has continued to portray itself as a developing nation working with others in the developing world on a common agenda. As a state, it prefers a multipolar world rather than one of superpower relations. It has, moreover, sought to cultivate friends in places that other major powers have forgotten or have been unable to engage with for political reasons. Such considerations have not posed an obstacle for the PRC, whose policy has been one of noninterference in other nations' domestic affairs, even in times of strife and violence.

The rare-earth crisis of 2010 enabled China's critics to sound the alarm and to promote panic with pronouncements of a dramatic shift in China's international policies and outlook. Its state competitors—the larger industrial nations—sought to collaborate with one another to challenge China's position vis-à-vis its resource-management policies and to rush to Japan's aid so that the territorial dispute would not escalate further. Extensive political maneuvering took place to put China on notice, at least until this particular crisis subsided. Given the level of international preoccupation, even more surprising is that prior to this first geopolitical and market crisis in 2010, these essential rare-earth inputs for green-energy, high-tech, and military applications had for the most part been bypassed in analyses of development strategies, especially by the strongest actors, namely the United States, the European Union, and Japan. Much research into the properties and possible applications of rare earths had been reported in the scientific journals, but these data had not factored into the scholarly discourse on economic statecraft and resource competition and the possible ripple effects in the international arena.

For almost two years, the rare-earth crisis made headlines and produced widespread reactions. Governments chastised China for using quotas and taxation of its rare-earth exports. Private-sector interest in investing in the few mature rare-earth mining projects outside China grew as prices continued their parabolic rise. Scientific cooperation between the EU, Japan, and the United States flourished, as they looked for ways to substitute the critical elements and innovate out of their use. Finally, the United States, the EU, and Japan disputed China's policies at the World Trade Organization, asking the PRC to scrap all its restrictions. Eventually, the market began to correct itself. Prices came down significantly and have even plummeted in the cases of the more abundant rare earths. The world has once again shifted its attention to other more "pressing" matters. Many claimed that the crisis itself had been hype, and that rare earths were in fact so abundant that the world would never be wanting.[18] Others sought to portray the international incident of the "unofficial" embargo as a negligible occurrence. China, in their view, never intended to overplay its hand and flex its muscles. They could not find any change in China's foreign policy tone.[19] Some analysts agreed that China had merely tried to address and attempt to reverse the significant damage that is done during both the mining and the refinement process. This had growing ramifications on the environment and on agricultural production, and had, moreover, fueled public discontent over health concerns. At an institutional level, the WTO eventually tried the case and China lost its appeal.[20] The fact that China today participates in the WTO is, for many analysts, of itself an indication that China increasingly understands the rules and resolves disputes within the organizational confines.

Accordingly, it has been argued that China's attempt to use its near-monopoly position has not produced any lasting benefits for itself, prompting the world to return to business as usual. The global world recession, continuing illegal smuggling of rare earths from China, and the considerable stockpiling that took place during the crisis have further dampened the prospects of a short-term rare-earth price recovery and given ammunition to those who have simply moved on. There is a widespread sense among analysts and policymakers that the events of 2010 were a mere passing blip on the screen.

The question I examine in this book, however, is the following: Was the rare-earth crisis really just a temporary glitch in the international system, without subsequent political and economic repercussions? I think not. If one takes a closer look at the overall situation, one can't but notice that China maintains its near-monopoly position. Moreover, the current low prices of all commodities, including energy materials in particular and rare earths, mean that the mines outside China that drew initial investment because they had the most mature prospects of providing rare earths are now straddled with debt and face difficulties staying afloat. Both Molycorp in the United States and Lynas in Australia saw their stock prices drop from 74.22[21] dollars and 2.65[22] dollars, respectively, in 2011, to 0.36 cents and 0.04 cents, respectively, in July 2015. Molycorp, moreover, filed for bankruptcy in June 2015.[23]

Since 2010, China has worked to centralize its production system to control smuggling and to introduce environmental standards that had been lacking, especially in small, unregulated extraction and processing facilities. At the apogee of the crisis many companies chose to collaborate more closely with China, some even moved production there, to secure unobstructed access to rare earths. This development helped advance China's plan to add value to its economy from the production of these vital elements, transforming itself from a mere exporter of raw materials to a producer of final applications. Furthermore, the PRC has retained its expertise and superiority in research and development in both metallurgy and the development of new applications. Although China agreed to abide by the WTO ruling, between the complaint and the final verdict, it had sufficient time to achieve its own interim aims without overtly demonstrating a disregard for its membership obligations. So if China seems to have lost the battle, it may not have lost the war. None of the ambitious bills that were drafted by the US Congress in response to this crisis ever made it into law, nor has the scientific community been able to substitute and innovate out of the use of rare earths in such a way that a renewed crisis would have no impact. Recycling efforts remain difficult and are not financially viable, especially given the current prices of the elements.

The rare-earth crisis is not simply a trade dispute, however. Nor is it simply an economic equation. It cannot be viewed only through the lens of the industrial considerations for defense, high-tech, and green applications. It also raises questions about China's use of economic statecraft[24] and points to outcomes of growing resource competition. It serves as a case study that reflects efforts by a "strategic" commodity producer to maintain and secure market share. Furthermore, the story of rare earths is one of technological innovation, economic competition, and dominance and the possibilities of conflict and cooperation. This complexity is what policymakers in international affairs must take under consideration as they analyze events to develop their responses and strategies in an increasingly globalized world. Given the complexity of such a task today, it becomes imperative for particular irrefutable facts to not be overlooked in haste.

First, with the world population set to grow to nine billion by 2050, competition over resources has been predicted to grow,[25] especially with regard to energy resources. The rise of Brazil, Russia, India, and China, and other countries in the developing world means that millions of people will be joining the ranks of the middle class and will want to be part of what we think of as a modern way of life—a life that is not only energy intensive but also characterized by high consumption. Second, there is evidence of a growing frequency of resource nationalism rearing its head in a number of instances.[26] Countries have been looking for better ways to achieve higher prices for their raw materials and, most importantly, have been actively seeking to diversify their economies so that they are not wholly dependent on rents and resource exports, in an attempt to shield them from exposure to the cyclical boom and bust cycles of commodity trade. Third, nanotechnologies and the growing complexity of applications now require an unprecedented range of materials, all of which come from the far reaches of the world. Rare earths are enablers and will continue to be critical to the diverse menu of ingredients that modern technologies require to function. Fourth, there are today countries that hold a strong position in the production and export of a vital commodity, which are seeking to use their current positions to defend their own market shares. They have attempted to both preempt investment in the extraction of these materials from areas of new discovery and to block, if possible, enhanced production by certain perceived rivals whose additional product could dilute their market position.[27]

In China's case, its dominant control over the rare-earth elements and its ability to manipulate the market has allowed it to maintain its market share by discouraging, canceling, or buying up investments in mining projects outside its borders. One of its underlying objectives could well be to continue its stranglehold of the high-tech industry, the production of renewables, clean technologies, and defense systems. Furthermore, China, in many instances, also controls the supply chain from mine to market and therefore controls the product itself (e.g., the production of magnets and monitors). This kind of power could again lead to production disruptions if China were to decide to limit access to rare earths in the future.

These facts, though significant, seem to have been overlooked or underestimated in the assessment of what the rare-earth crisis meant on a global scale and its potential to recur in the future. At the very least, it underscores the increasing lack of medium- and long-term planning by political actors, who, under the pressure of the here and now, have exhibited knee-jerk reactions to the challenges of the moment. It is not only the political system that is so short-sighted. The rare-earth crisis proved that those businesses that were willing to pay a premium for REEs produced outside China during the height of the market are now unwilling to support the few alternative sources of product that exist unless the prices are equal or lower than those in the PRC. Furthermore, the direct response of some companies in the critical years 2010 and 2011 was to move production of their products to the PRC to ensure lower prices for rare earths and uninterrupted access to them.

In this book, therefore, I aim to present a more nuanced and holistic interpretation of what I consider a significant and distinctive paradigm of resource competition, one that offers an important and fruitful case study for established theories of international relations. I also intend to explain how the events that made the rare-earth crisis prominent in the world community are open to a variety of interpretations that can lead to significantly different assessments of their significance and impact in the long term. While some analysts and policymakers securitized the problem and raised a battle cry to stop an aggressive China, for example, others argued that the market would find solutions to resource and other trade disputes, insisting that there was never a cause for alarm. These kinds of conflicting and opposing analyses can cloud the issue for governments and practitioners of international affairs, when there are, in fact, many lessons to be learned from the crisis that should not be brushed aside because the tempest has subsided. At the very least, this book, by hanging a lantern on the problem, may prompt policymakers to reassess how effectively they have prepared for the future. As global challenges continue to mount, it should at least be acknowledged that policies drafted based on an "out of sight out of mind" approach will not provide the necessary alternatives and solutions.

To set the stage for my discussion of these larger questions provoked by China's dominance over the supply of rare earths and the world's growing dependency on them, chapter 1 begins with the general theoretical debate concerning resource competition, mineral scarcity, and economic statecraft and their impact on international affairs. Next, because of the increasing geopolitical and strategic importance of rare earths, it is helpful to examine the history and mechanisms by which they have come to occupy such a vital position in critical industries.[28] In chapter 2, I detail their increasingly widespread use, their impact on technological change, the supply/demand aspects of the industry, and the limited possibilities for effective substitution and recycling.[29]

In chapter 3, as part of this larger examination of the international political and strategic issues raised by China's current monopoly of rare earths, I look at two relevant historical examples of resources that became strategically important—salt and oil—exploring analogies for the current rare-earth crisis and suggesting how such parallels help us to better understand key variables in the present case. I then present a detailed account, in chapter 4, of how China came to dominate the industry by design and how other industrial powers have reacted to China's stranglehold on these elements. Just as Deng Xiaoping and his successors had foreseen, as enablers, rare earths will remain indispensable in many vital applications of modern life and, for this reason, will continue to play their part in China's design of its long-term grand strategy, affecting global geopolitical relations in ways with which China's potential rivals will need to come to terms.

# Resource Competition, Mineral Scarcity, and Economic Statecraft

The single biggest development in international affairs since the Cold War has been China's meteoric rise. It is primarily through its economic transformation and success that China has gained political importance—a fact that explains why its actions are now under widespread scrutiny. China's role as the "economic powerhouse," moreover, has transformed the global economy and generated important debates on issues such as deindustrialization, the loss of jobs in the West, the management of the global economy through international institutions, industrial policy, the role of state-owned enterprises, and, in the case, of rare earths, growing resource competition and how it impacts other industrial nations.[1] Its economic clout is principally what has allowed China to pursue its strategic objectives on a global scale and has given rise to a debate about a reshuffle in global power relations that profoundly affects relations among states.

China's resource management policies over rare earths, a seemingly straightforward economic issue, spilled over dramatically into the political sphere the moment that a confrontation with Japan over disputed waters came into the equation. It was not the deep cuts in quotas that triggered international outcry. It was rather the use of economic statecraft—in the form of a temporary (albeit unofficial) disruption of access to supplies of an important resource—that raised international alarms. This decision influenced how other nations began to read China's handling of its role as the main supplier of rare earths. Under the lens of political scrutiny, its stranglehold on the rare-earth market came to be interpreted as a power play because of China's use of known tools of economic statecraft to achieve particular geopolitical and strategic ends.

## RESOURCE COMPETITION

Resources, renewable and nonrenewable alike, have always been at the heart of our construction of the man-made environment. Throughout human history, there have been a number of resources able to break civilizations and empires. At different times and for different reasons, bronze, iron, silver, gold, salt, and coal have been at the epicenter of economic development, dominance, and technological innovation, and conflicts and full-fledged wars have broken out in order to obtain them. New trade routes were created, for instance, and the New World was discovered in their pursuit; colonialism arguably developed as a way for European nations to exert control over resources.[2]

Since the Industrial Revolution in particular, as the world has been on a trajectory of exponential growth, resources have been the essential inputs in the world economy and at the epicenter of technological developments. For these reasons, they have featured prominently in the discourse of "resource wars," a term coined and used extensively starting in the 1980s in relation to US and Soviet efforts to control fuel and minerals in peripheral countries that belonged to each other's sphere of influence.[3] General structural theories, furthermore, explain interstate resource wars as a rational calculation of a state's interests in a situation where there is a fixed or shrinking pie of natural resources.[4] Clearly, governments across the globe have long understood their intrinsic value for economic growth, security, and power. Accordingly, over the centuries, they have devoted both resources and considerable effort to protecting and securing uninterrupted access to them for their own economic development.

As military weapons have grown increasingly powerful and destructive, however, direct conflict over resources—though it remains an option—has been complemented by significant attempts at interstate cooperation. The building of economic and political ties through bilateral and multilateral cooperation as well as global trade took center stage to ensure peaceful yet uninterrupted access to vital economic inputs. Supply flows became a central preoccupation of the developed world, especially with respect to the possible vulnerabilities of the most critical inputs, such as oil.[5]

Today, nations continue to worry about access to resources, even more as it becomes clearer that we are now witnessing a drastic change in the new pace of their extraction and exploitation. It is this change that has significantly altered the equation and is adding a new layer of global concern. According to John Tilton,

> Humankind has consumed more aluminum, copper, iron and steel, phosphate rock, diamonds, sulfur, coal, oil, natural gas, and even sand and gravel during the past century than all earlier centuries together. Moreover, the pace continues to accelerate, so that today the world annually produces and consumes nearly all mineral commodities at a record rate.[6]

Making the shift even more pronounced is the rapidly expanding "omnivorous diet of modern technology."[7] In their 2013 article in *Resources, Conservation and Recycling*, Aaron Greenfield and T. E. Graedel succinctly and aptly point out that "two centuries ago the diet of technology (the diversity of materials utilized) consisted largely of natural materials and a few metals. A century later, the diversity in the diet had expanded to perhaps a dozen materials in common use. In contrast, today's technology employs nearly every material in the periodic table."[8] Our modern appetite for new technologies and new applications appears to be insatiable and requires resources from all around the globe. As a result, the global demand for the resources that are pivotal for the functioning of modern-day economies[9] is the highest it has ever been.[10]

Given our dependence on these inputs, the possibility of resource scarcity now poses a major concern to nations across the globe. Aggravating this worry is the fact that the world population is poised to reach nine billion, and the growth of the middle class in developing nations is requiring manufactured goods and energy at unprecedented levels.[11] Price volatility of inputs is a source of anxiety as well, especially because of the significant increase in industrial demand.[12] According to a report prepared for the EU Commission in 2012, historic trends indicate that prices, especially since 1998, have been steadily increasing. The aggregate price index included in the report in fact showed that on average real prices increased by more than 300% between 1998 and 2011. Moreover, "a big jump occurred between 1998 and 2000 when prices more than doubled (i.e., a 125% increase in real prices), but even after 2000, prices have increased further by almost 6% per annum in real terms."[13]

The potential of resource scarcity, therefore, is a matter of concern around the world not only for governments, but also for companies. Nonrenewable resources are essential because our modern economy is dependent on their availability. This means that diverse actors have begun to look for ways to address the problem of growing demand and the possibility of confronting a steeper supply curve. They must, therefore, invest in new discoveries and in securing access to existing reserves. The threat of shortages and growing competition has also triggered the need to acquire mining rights from poor and developing countries and commodity producing firms[14]—a policy that poses a whole other set of security and geopolitical issues.

It is important to note, however, that not all scholars and analysts share this worry about the long-run availability of mineral resources. There are those who believe that the market, in conjunction with tailor-made public policies, can offset the threat. Prices will rise, they argue, sparking further exploration, research and development (R & D), recycling, and substitution; consumers, responding to higher prices, will shift away from these resources to other alternatives. The question of whether the market can be a panacea in situations of resource scarcity, taking into account the significance of an unsustainable rate of consumption of the vital resources we use to power, run, and grow our modern world, provides an important backdrop for understanding the trajectory of rare earths and their impact.

An overview of different notions of "resource scarcity" and in particular "mineral scarcity" will permit a better understanding of how essential resources can impact global relations, triggering deeper cooperation or rivalry and conflict.

## WHAT IS MINERAL SCARCITY?

Although minerals can be found throughout the earth's crust, they are characterized as a resource only once they have become useful as inputs for human applications and technologies. At that point, concern arises about their availability and the degree of their abundance, both physical and economic. These two concerns, availability and degree of abundance, usher in the concept of scarcity. Minerals are viewed as scarce for two principal reasons: because of either stock depletion or flow disruption.

Thus far, stock depletion of any particular mineral has yet to occur, but flow disruption is more common and has happened repeatedly. If the flow of the materials is inadequate and demand is not satisfied, then markets react by driving prices upward. Flow disruption can occur for of a variety of reasons, such as a temporary halt in production or a sudden increase in demand that far exceeds production. It can also result from policy choices made by governments, such as embargoes.

Both the markets and industries can address resource scarcity by using a wide range of options and tools available to them. Investment in new exploration, for example, may begin to boost production; companies can look at material substitution or invest in R & D to innovate out of the use of particular minerals or, at the very least, reduce their dependence on individual inputs. Other options include the implementation of recycling schemes and the introduction of more efficient production processes that can limit the impacts of scarcity. Sometimes industry responds by producing lower-grade ores or by extracting materials from marginally economically viable deposits. Governments also have a vital role to play in these situations. Furthermore, they can intervene with policy instruments aimed at helping to stabilize prices. Some countries, for example, stockpile critical minerals that they can release if there is a shortage. Some additional financial instruments that are available include initiating price controls, subsidizing consumers, or giving tax deductions to industry. Governments also can respond by designating funds for research and development to encourage innovative solutions.

Although, as mentioned previously, we have yet to experience stock depletion, the potential physical scarcity of minerals[15] has been hotly debated among economists and geologists. There are two perspectives on the physical scarcity of minerals: the fixed-stock and opportunity-cost paradigms,[16] and the debate between them has become increasingly polarized. Resource exhaustion[17] has for centuries loomed as a real possibility and fear of it dates back to Thomas Malthus, the eighteenth-century economist.[18] In the 1970s, the oil shocks reanimated the Malthusian paradigm, only to be put aside in the 1980s after prices came down for both energy resources and minerals.[19] The 1990s ushered in new questions about depletion, coupled with the dangers of environmental degradation.[20]

John Tilton offers two examples that show the level and nature of this polarized debate.[21] He cites both a geologist (Kesler) and an economist (Simon). Kesler points to increasing physical depletion and the environmental impact of extraction:

> At the end of the twentieth century, we are faced with two closely related threats. First, there is the increasing rate at which we are consuming mineral resources, the basic materials on which civilization depends. Although we have not yet experienced global mineral shortages, they are on the horizon. Second, there is the growing pollution caused by the extraction and consumption of mineral resources, which threatens to make the earth's surface uninhabitable. We may well ponder which of these will first limit the continued improvement of our standard of living.[22]

For Simon, the question of scarcity is one of economics.

> People have since antiquity worried about running out of natural resources- flint, game animals, what-have-you. Yet, amazingly all

the historical evidence shows that raw materials—all of them—have become less scarce rather than more . . . Natural resource scarcity— as measured by the economically meaningful indicator of cost or price—has been decreasing rather than increasing in the long run for all raw materials, with only temporary and local exceptions.[23]

The fixed-stock paradigm assumes that the supply of any mineral is finite and therefore exhaustible. Supply diminishes with use as demand continues to grow. Although those who adhere to this school of thought understand that technology and price increases have already led to new discoveries, they are also wary of the rapidity of unchecked technological leaps that can readily lead to potential calamities.[24] The continued extraction of minerals ultimately leads to scarcity, since in their view, measures of physical scarcity are associated with the fixed-stock paradigm.

The opportunity-cost paradigm, however, views the issue of fixed stock as almost irrelevant. In this view, resources are not exhausted by their use; they remain, and depending on the price, can always be recycled and reused. When a mineral resource becomes less available, its price rises. Such lack of availability naturally leads to further exploration, new findings, and new technologies. In 1963, Barnett and Morse argued that new technology and other developments had effectively reduced costs, offsetting the cost-increasing effects of depletion.[25] Their data showed that for agriculture and minerals, price and production costs had fallen or remained constant during the period from 1870 to 1957. In their analysis, technology was instrumental in producing substitutes for scarce resources. It had decreased the extraction costs of minerals and, in that way, was able to expand the size of economic reserves. Those who favor this kind of opportunity-cost paradigm believe that physical exhaustion would, in any event, not occur before the price would rise exponentially. At that point, recycling, reuse, and resource management would kick in and become more affordable alternatives. Ultimately, outcomes would depend on what people are willing to pay for the particular mineral (opportunity cost).

Price, however, may not be the most reliable indicator of mineral scarcity, since many costs, such as environmental and social ones, are today not reflected in the market.[26] However, cornucopians, such as Boserup, Simon, Lomborg, Juul, and Mortimore,[27] further argued that scarcity would rarely occur because three important factors provide solutions—namely, international trade, technological innovation, and substitution. Briefly, cornucopians argue that the increasing demand for resources can be met by continued technological advances. For this reason, population growth is not viewed as a constraint. Resources will therefore suffice in conditions of growth. These views have been contentious and drawn much criticism. One such critique comes from Thomas Homer-Dixon, who, in his article "Cornucopians and Neo-Malthusians,"[28] argues that cornucopians overlook six important factors that make today's scarcities different from what the world has experienced in the past.

According to Homer-Dixon, the growing uncertainty facing policymakers and economic actors largely owes to the fact that current scarcities are both multiple and developing at a faster pace. Consequently, this reduces the effects of economic, technological, or social adjustment. Moreover, even though today consumption rates are unprecedented, there is no indication that companies or consumers are prepared to alter their practices or behavior. In fact, quite the opposite seems to be the case. Furthermore, Homer-Dixon points out that the free market is not an adequate tool for measuring scarcity, "especially when applied to such resources as sea productivity and a good climate."[29] What is more, though market responses to scarcity may work better in richer countries than in poorer ones, it is the latter that face the most pressing environmental issues. He expresses deep doubt that in the future, humanity can continue to innovate itself out of scarcity effectively and in time. The challenges facing us today depend on confronting extremely complex environmental and social systems whose intricacies will not allow for the kinds of speedy solutions envisioned by cornucopians.[30]

There is yet another group of scholars who assert that metal supplies can be discussed from both a physical and an economic perspective. Gordon, Bertram, and Graedel,[31] for instance, examine the case of copper. They bracket both scarcity and "running short," and instead attempt to quantify possible resource challenges by examining issues such as absolute amounts, rates of use, rates of loss, and developmental scenarios.[32] Their basic argument about the sustainability of metal resources is that "an economy driven by cultural and political forces to dependence on an ever-increasing flow of material goods through use providing services to end-of-life and into waste is ultimately unsustainable."[33] At the same time, they invoke the writings of Thomas Princen, who argues that eventually "sufficiency will replace growth as the economic norm."[34]

In a similar vein, Markus A. Reuter and Antoinette van Schaik, in their chapter "Transforming the Recovery and Recycling of Nonrenewable Resources," claim:

> Minimizing the losses of nonrenewable resources requires creating the closest possible approach to a truly circular use of resources, a goal emphasized by Moriguchi as crucial to sustainability. Achieving this goal will require harmonious interaction among all actors in the resource cycle: the mining industry, metallurgical processors, original equipment manufacturers (OEMs), nongovernmental organizations, ecologists/environmentalists, legislators, and consumers (to name a few).[35]

Mineral scarcity can also result from a disruption in the chain of supply and demand, known as "situational scarcity." There are a number of reasons such disruptions might occur. Restrictions in the mineral supply can be a result of a significant increase in demand, thin markets, concentration of production, production predominantly as a byproduct, and lack of available old scrap for recycling or of the infrastructure required for recycling. Over the long-term, continued investment in the extraction and processing of minerals is a key factor, along with research and educating geologists, engineers, and miners to avoid possible supply disruptions. With regard to the repercussions import dependence may cause, the 2008 National Research Council report "Minerals, Critical Minerals, and the US Economy" underscored that vulnerability arises when, for example, supply is concentrated in "one or a small number of exporting nations with high political risk or in a nation with such significant growth in internal demand that exported minerals may be redirected toward internal, domestic use."[36]

Other forms of scarcity[37] can be provoked by government actions when policymakers resort to the techniques of economic statecraft. Embargoes of minerals are frequently a technique of choice to help push up the price.[38] Another tool can be the decision to limit exports or push for the renegotiation of a rent distribution. This causes disruption in the flow of the commodity. There have historically been interruptions of shipments during times of war. Robert Mandel labeled this kind of threat "the 'strategic minerals threat' and [noted] that even if it may only be a perception, it carries the strength of a weapon if applied."[39]

Examples of scarcity provoked by "government" actions include Allied attempts to curtail shipments of oil to the Axis countries during World War II. Moreover, during the Cold War, the United States and its allies feared that the USSR would use its sphere of influence and proxy states to restrict mineral supplies to the Western Bloc.[40] Another such example was the prolonged embargo[41] by the United States, the member states of the European Union, Japan, the Republic of Korea, Canada, Australia, Norway, Switzerland, and others against Iran[42] that was meant to serve as a "strong, inter-locking matrix of measures relating to Iran's nuclear, missile, energy, shipping, transportation, and financial sectors."[43] Among the reasons for the sanctions cited by the US State Department was to "induce Iran to engage constructively, through discussions with the United States, China, France, Germany, the United Kingdom, and Russia in the 'E3+3 process,' so as to fulfill its nonproliferation obligations."[44]

Today, there are new push factors that could lead governments to consider taking actions that may create scarcity conditions in the world markets. These include rapid population increases, the rapid economic growth of the developing world, and the complexity of applications requiring many more inputs, as well as the growing climate crisis. One way to respond to these push factors is to use a technique, again associated with economic statecraft, whereby nations decide to limit exports in order to fulfill their own domestic needs instead of participating in the open market, seeking to gain greater control of their own vital assets. In this case, they may choose to use different tactics and policies to enhance their control vis-à-vis their vital natural resources: increased taxation, nationalization, forced partnerships, and new regulation are among the usual options. This kind of behavior has been termed "resource nationalism." There have been a number of recent cases. To cite just a few: in 2006 Bolivia nationalized its oil and gas industry; Chad created a national oil company and expelled both Chevron and Petronas in August of that same year; and Venezuela[45] mandated that foreign-owned assets be sold, a decision that effectively pressured Exxon and Conoco Phillips to leave in 2007.[46] Critical metals and energy sources are at the heart of this particular debate. In the case of China and rare earths, the 2010 restriction on rare-earth element (REE) exports was repeatedly "labeled" as the implementation of just such a policy tool, reflecting a growing "resource nationalism"[47] on the part of the People's Republic of China (PRC). By the same token, another possible course of action in response to the aforementioned push factors that may create scarcity conditions in the world markets is for governments or companies (state and private) to seek to purchase mines in developing countries and enter long-term agreements for mineral rights, thereby exercising exclusive control of those resources. China is already making these kinds of strategic acquisitions, primarily in Africa and Latin America, further indicating to some that China is implementing a strategy of resource nationalism.

Political instability and internal warfare can also disrupt the flows of minerals to the market. The political instability caused by industrial nations competing over access to coveted resources contributes to the creation of more failed states and protracted civil wars, at the same time drawing in stronger states into growing disputes. Homer-Dixon summarized the threats to industrialized nations in his book *Environment Scarcity and Violence*:

Although this violence affects developing societies most, policymakers and citizens in the industrialized world ignore it at their peril. It can harm rich countries' national interests by threatening their trade and economic relations, entangling them in complex humanitarian

emergencies, provoking distress migrations, and destabilizing pivotal countries in the developing world.[48]

In the cases of poor but mineral-rich nations, abundance leading to greed
often undermines their stability and welfare. Becoming exporters of natural resources can weaken their motivation to take the time to develop
multiple sectors of the entire economy. They mostly rely, instead, on the
revenues from the sale of their valuable resources.[49] Their biggest assets
then become a kind of "resource curse,"[50] which ultimately undermines
their ability to function successfully as states.

By the same token, a variety of factors can lead to grievances among
local communities, which typically do not benefit from the extraction
and sale of valuable minerals. In addition to dislocation, there is often
the threat of environmental degradation of their land and water because
of the inadequate or complete absence of environmental oversight and
regulation.[51] Other factors are unfair working conditions and low com-
pensation. In Indonesia, for example, the citizens of Sulawesi complained
that Newmont Mining had been dumping tailings at sea, contaminating
the Bay of Bengal with arsenic and mercury.[52] In Ghana, the gold mines
led to the resettlement of the local population. Farmers were not compen-
sated, and violent removal was also considered to ensure their relocation.[53]
Accusations of environmental degradations raged in Malaysia in response
to the building of a rare-earth processing plant by the Lynas Corporation,
provoking opposition, disruptions, and unrest. When these and other
grievances[54] among indigenous populations cannot be or are not properly addressed by authorities in unstable nations with weak and nontransparent governments, civil wars often break out. Armed conflict and civil
strife can also result from power struggles to control the mineral wealth
by local elites. Similarly, some poor-yet-resource-rich nations have suf-
fered through outside military interventions to ensure that already signed
agreements are fulfilled and that access to their minerals is maintained.[55]

There are, nonetheless, other possible responses to mineral scarcity
that lie within the sphere of cooperation. Agreements over minerals
have ranged from gathering and sharing information, stockpiling, mine

production (exploration, development, and extraction), and, most nota-
bly, trade. Governments, industries, and social groups can cooperate to
achieve different outcomes with respect to minerals. This can take place
between states, regions, and companies, and institutional frameworks are
pivotal in facilitating cooperation.

There are a number of drivers of cooperation: international trade agreements, such as free-trade agreements; memorandums of understanding;
joint workshops, such as those held by the EU, the United States, and
Japan as part of their response strategy to the rare-earth crisis; corporate
social responsibility (CSR) policies; industry associations, such as mining
organizations, that promote industry interests; and business and community cooperation to ensure peaceful coexistence, as well as to offset environmental impacts. A number of international organizations have been
formed to help deal with potential armed conflicts over valuable commodities and limited resources. These facilitate and help structure collaboration, cooperation, and agreements between nations. The EU, the
United Nations, the World Trade Organization (WTO), the Association
of Southeast Asian Nations (ASEAN), the UN Intergovernmental Panel
on Climate Change (IPCC), and others are all there to help work out differences and reach common understandings.

More often than not, potential conflicts have been resolved through
political negotiation, pressure, and the free market. International collaboration and cooperation is thus recognized as an efficient way to seek
solutions and avoid unnecessary military conflicts. Nonetheless, there are
many reasons why such cooperation may stall, fail, or just fall by the way-
side. Market conditions and political events, local or international, may
intercede. We must, furthermore, bear in mind that these institutions,
set up by the United States and its European allies, are now changing as
well, as other countries rise to take their place among the most powerful
nations in the world system. A power shift is emerging that also affects the
terms and nature of cooperation, and this certainly contributes to power
tensions.

By way of a preliminary conclusion: resource scarcity does not necessarily mean the physical depletion of a mineral. Many other factors can

come into play, all helping to intensify the competition over access to particular resources and provoking various outcomes.

## ECONOMIC STATECRAFT

We have seen how resource scarcity, whether physical or provoked by flow disruption, poses a series of challenges and concerns for policymakers and industry. Arguably, today's multiple scarcities of critical resources have fueled uncertainty in the minds of policymakers and economic actors. Policymakers have many tools in their toolboxes with which they can attempt to address impaired access to needed resources. Direct, even military confrontation, is one possible tactic, but attempts at cooperation—diplomatic, scientific, and economic—are also possible. The underlying reason for either course of action is the quest for security and access to continued prosperity through economic growth and development.

In search of solutions, countries resort to "statecraft," which is essentially another way of describing the art of conducting state affairs, in particular, the strategies used in pursuit of foreign policy goals. When an issue of resource scarcity emerges, statecraft does not limit itself to techniques that may lead to either diplomacy or war. This is particularly important to consider in a nuclear world that begs for alternative solutions to military confrontation. In a time when all-out generalized war has become unthinkable, economic statecraft becomes even more salient.

"Economic statecraft" is what provides policymakers with something that they need above all else when strategizing—namely, options. It offers them alternative courses of action. Policymakers require a range of solutions that they can compare in order to weigh the pros and cons and estimate the costs and benefits of the different options that are brought to the table. As David Baldwin explains in his seminal book *Economic Statecraft*, "The essence of foreign policy is its purposive behavior," meaning that it is designed to take action to achieve a goal or an aim. These goals may not only involve multiple targets, but they may also carry varying degrees of significance. In foreign policy, nothing remains static, and situations are re-evaluated and reprioritized according to state actors' most current assessment and understanding of a situation.[56]

Different instruments of economic statecraft are often used by nations seeking to achieve a wide range of goals: embargoes, boycotts, tariff increases, tariff discrimination, withdrawal of "most-favored nation" treatment, blacklists, quotas (import or export), license denial (import or export), dumping, and preclusive buys are all used regularly in trade relations. With respect to capital, the available options can include the freezing of assets, suspension of aid, control of imports or exports, expropriation, taxation, and withholding the payment of dues to international organizations and finally threatening any of the above actions.[57] Some of the most common goals of economic statecraft are to weaken, strengthen, or even change the leadership or the political system of another state. Alternatively, the objective might be to acquire or attract allies from another state. Sometimes the aim is to stop an ongoing war or to affect the tariff policy or to change the rate of growth of another state.

There is nothing novel in these techniques. Economic statecraft has been used in the past and will continue to be used in the future by states aiming to achieve a wide array of foreign policy goals. In fact, there has been an increase in its use since the end of the Cold War. The United States, for example, has frequently employed economic sanctions as an instrument of foreign policy.[58] Other countries and the United Nations Security Council have resorted to the use of sanctions as well.

Free trade is itself a technique of economic statecraft, and one that has since its inception been very effective. It was one of the main instruments the United States used after the Second World War, not only to help Western Europe and Japan recover from devastation and to develop the poorer countries of the South, but also to help strengthen its military alliances and to create new markets for its own exports, as well as to help build a new sense of peace and security after years of conflict and displacement.

Such a framework of economic statecraft helps to shed light on how China and the United States, the EU, and Japan chose to address and offset what they perceived as a growing threat of resource scarcity during the rare-earth crisis of 2010. At the very least, economic statecraft theory supports the argument that the rare-earth crisis of 2010 did in fact mobilize a series of actors, who chose a particular set of policies from an array of techniques that they thought would restore their access to a vital resource, taking China to the WTO, for instance. Furthermore, it allowed the Chinese to preserve (through export quotas and taxation) what they themselves had evaluated as a strategic resource, boost their economy by adding value to the supply chain, address internal pollution concerns, maintain their supply dominance, and parlay this strength, at least on the surface,[59] additionally into a response to a territorial issue.

In the scholarly literature, there are those who argue that the overall use of economic statecraft has perhaps not proven to be as effective as is widely supposed, especially in today's interconnected world. A look at China's attempts to use economic statecraft following the 2008 global financial crisis, for example, not only allows for different interpretations of its success, but also provides a window into a period of Chinese policy that has been characterized as increasingly assertive.[60] The 2008 crisis and the global financial instability it caused did not leave China indifferent but prompted a varied set of responses. First, there were attempts to influence and even dictate new terms for US financial policy via a series of statements and threats aimed at assuaging China's internal insecurities growing out of the fear of a possible sudden economic slowdown and what it would mean for the government's legitimacy. Examples of the warnings emanating from the Chinese side include statements, such as that of the prime minister Wen Jiabao, in March 2009, who voiced China's concern over the safety of its US assets, which amounted to a $1 trillion investment in American government debt. In support of this strategy, the head of the People's Bank of China, Zhou Xiaochuan, further suggested reforms in the international monetary system.[61] At the same time, influential actors in China[62] read the crisis as a clear indication of American decline that provided an opportunity for China to now use its clout and project its newfound strength in the international arena in a more decisive manner.

The threats and rhetoric employed at the time, however, did not lead to significant changes or policy concessions. They did not even deter the

United States in 2009 from joining the EU in filing a case with the WTO challenging China's right to restrict bauxite, coke, magnesium, manganese, and zinc exports.[63] While the assertive language used by the Chinese led the United States to issue assurances about the safety of American bonds, the US government did not hesitate to proceed with fiscal and monetary expansion to counter the crisis at home.

While the use of—in this case—financial statecraft on the part of China may not have resulted in what can be interpreted as major policy shifts, it did seem to indicate a change in Chinese assertiveness in the international arena. It also provided the Chinese government with a wider range of options in dealing with the repercussions of the crisis on their own economy. The PRC allowed the renminbi to depreciate, for example, and gave tax incentives to its exporting industries without raising much complaint from the United States. Some may also argue that China's stance in that period led the new Obama administration to initially soften its diplomatic tone to Beijing, which in some domestic Chinese circles was interpreted as a sign of recognition of the effectiveness of China's demonstrated assertiveness.[64] The framing of China's diplomacy and foreign policy objectives undoubtedly colors the ways in which governments, analysts, media, and the wider public perceive or report on the PRC's actions and intentions. As China's influence and impact on the global economy and international institutions grow, so does the perception that it is now intentionally demonstrating this growing assertiveness.

Some scholars, however, argue that there isn't enough of an indication that this is in fact the case. With respect to the question of China's overt "assertiveness" and use of economic statecraft in the case of rare earths, Alastair Iain Johnston,[65] for instance, recently looked at a series of incidents in an attempt to demonstrate that China's reactions have often been misinterpreted or are, instead, a product of interbureaucratic conflict and poor coordination. He examines the 2010 "unofficial" rare earths embargo against Japan in defense of this claim. Johnston's data is not detailed enough to reflect the exact nature of the rare earth shipments reaching different Japanese ports during the critical period in question. In one important instance, however, he does provide data of yttrium shipments that indicate an across-the-board decline of yttrium imports from August to October of that year which undercut his main claim. The particular data are important because rare earths are not all equally valuable. Yttrium is one such rare earth, used extensively by the lighting industry. It is much more valuable and rare, for example, in comparison to the more plentiful and readily available cerium oxides that Johnston highlights in his analysis. In fact, as I will argue, the distribution between heavy and light rare earths is crucial for assessing their criticality for particular industries.

Be that as it may and though the curtailment of shipments may have been short-lived—because Japan released the Chinese captain—its impact on the market lasted much longer because it overlapped with the sharp reduction in export quotas. Accordingly, however one interprets Johnston's data, there is no denying that in terms of global political perception, the Chinese response further underscored the growing tension between the two nations and the potential range of spillover effects. One particular implication of the dispute between Japan and China is that it draws in the United States. The United States recognizes the administrative control of the islands by Japan, which as a result fall under the US-Japan Defense Treaty.[66] With hindsight, each new attempt by China to weaken Japan's sole claim to these islands lends credibility to Japanese fears that, with respect to its sovereignty, China's rise may not be entirely benign.

The major industrial powers that took the lead in opposing China's rare-earth policies during the 2010 crisis were influenced by varying concerns and perspectives that colored their interpretations of what China was attempting to achieve by its actions. It is true that after the initial round of diplomatic pronouncements, industrial countries gradually downplayed their geopolitical concerns over the matter. Even so, the United States, the EU, and Japan did share the immediate common trade goal of repealing export quotas and taxation on the elements that were considered vital inputs for their industries. All three cooperated, using techniques of economic statecraft, to confront China and to return the market to normalcy. Underlying these clear intentions, however, the wider concerns about China's rise and its goals growing out of their respective post–World War II visions of the workings of the international order had an impact on their reading of the situation.

My primary aim here is to give an account of a crisis situation that erupted because of the *perceived* scarcity of vital resources and of the substantial geopolitical considerations this raises. By the same token, I hope to underscore the shortsightedness exhibited by many of the involved governments, industries, and analysts, whose interest rapidly waned as soon as the imminent geopolitical and economic/resource threat subsided. These actors have failed to follow up on the conditions that led to the crisis and the possibility of its recurring again more strongly in the future.

Reflecting on the implications of China's rise in the international order goes a long way in explaining the political and economic nuances of the rare-earth crisis. Today's world order is the product of two catastrophic world wars and the nuclear and ideological rivalry of the Cold War. China did not partake in the construction of this post–World War II world. Indeed, it did not re-enter the world stage in any significant way until the 1970s. In its absence, the world created organizations and institutions that reflected the political, economic, and strategic desires and insecurities of the most important and engaged players of the time, namely, the United States, the Soviet Union, and Western Europe.

After the Soviet collapse, however, the world briefly was left with one traditional hegemon, a very powerful United States. US hegemony did not last long because the European Union stepped up its commitment to world affairs and China rejoined the international community with a strong economy and a clear plan. Today, the world system reflects a distinct multipolarity, though the United States consistently remains the power that views and acts in the world in strategic terms and remains involved in its operation.[67]

Joseph Nye, offers an insight into this changing world structure in his book *Soft Power: The Means to Success in World Politics*. He claims that

> the agenda of world politics has become like a three-dimensional chess game in which one can win only by playing vertically as well as horizontally. On the top board of classic interstate military issues it makes sense to speak in traditional terms of unipolarity or hegemony. However, on the middle board of interstate economic issues,

the distribution of power is multipolar. The United States cannot obtain the outcomes it wants on trade, antitrust, or financial regulation issues without the agreement of the European Union, Japan, China, and others . . . And on the bottom board of transnational issues like terrorism, international crime, climate change, and the spread of infectious diseases, power is widely distributed and chaotically organized among state and non-state actors.[68]

This changing world structure has already led to fundamental strategic repositioning and a redefinition of key interests. The United States, for example, after decades of prioritizing Europe and the Middle East in terms of its strategic interests, has shown a significant shift toward the Asia-Pacific region.[69] Although the United States had never in effect "left" Asia, the Obama administration gave US engagement with the region a sharper and more explicit geostrategic frame, distancing itself perhaps from the Bush administration's policies in Iraq, Afghanistan, and the Middle East.[70] Although the George W. Bush White House had renewed its engagement with America's Asian allies, it did so primarily within the framework of countering terrorism at a global level in the aftermath of September 11 and the subsequent US declaration of the "war on terror."[71]

The Obama White House went on to build on many of the previous administration's initiatives—such as the strategic economic dialogue with China[72]—combining them with a number of innovative and proactive policies in the region. It signed the ASEAN Treaty of Amity and Cooperation and joined the East Asia Summit. What stood out, however, was the rhetorical frame of US engagement in Asia. When President Obama stated that the United States is "a Pacific power,"[73] and that it would "play a larger and long-term role in shaping this region and its future," it was to underscore what the Defense Department's January 2012 strategic-guidance document had highlighted, namely, that the United States' "economic and security interests are inextricably linked to developments in the arc extending from the Western Pacific and East Asia into the Indian Ocean region and South Asia."[74] A few days earlier, Hillary Clinton had given a speech, entitled "America's Pacific Century," that highlighted the nature of the administration's renewed embrace of the Asia-Pacific region. The tone of the announcement, though upbeat, clearly indicated that the United States considered as its role the "shaping [of] the architecture of the Pacific" in the twenty-first century.[75]

The pivot toward Asia echoed the emerging view that in the twenty-first century China has become the United States' principal economic competitor and political rival on a global scale.[76] China is at the center of US strategic thinking and planning.[77] Although the US economy is intrinsically tied to China through investment, consumption, and debt, underlying this level of cooperation is a mistrust and an ideological chasm in both their models of governance and their respective views of human rights. There is also a growing competition for allies, markets, and influence. This is why the United States decided to actively shift its attention and bolster its presence in the Asia-Pacific region, to reposition itself in order to deal with shifting power dynamics in the international arena. In this way, the United States signaled its intension to focus on existing alliances and to broaden cooperation networks to pursue and secure common interests.[78]

In its January 2012 report, for instance, the Department of Defense stated that the United States perceived China's continuous rise as having the potential to affect US interests in this economically dynamic region. While stressing America's interest in strengthening its bilateral ties with China, it simultaneously underlined that "the growth of China's military power must be accompanied by greater clarity of its strategic intentions in order to avoid causing friction in the region."[79] Furthermore, the report underscored the intention of the United States to "make the necessary investments to ensure that we [the United States] maintain regional access and the ability to operate freely in keeping with our treaty obligations and with international law. Working closely with our network of allies and partners, we [the United States] will continue to promote a rules-based international order that ensures underlying stability and encourages the peaceful rise of new powers, economic dynamism, and constructive defense cooperation."[80] Clearly, the shift toward Asia did not constitute a mere transfer of resources from one region to another. While it reflected a more systematic investment of strategic capital, time, and resources, it was also an unequivocal declaration of the US intention to sustain its global leadership, secure its interests, and advance its values. According to a report by the Aljazeera Center for Studies,

> this "rebalancing" is already under way, as is apparent from America's warming relationships with India and Vietnam, policy shift toward Burma (Myanmar), and planned deployment of 2,500 Marines at a new forward- staging base in Darwin, Australia, that is to serve as a launch pad for Southeast Asia. The United States is also building up forces on its territory of Guam, a key strategic enclave in the Pacific much like the British island of Diego Garcia in the Indian Ocean.[81]

There are those, such as Thomas J. Christensen, who argue that American engagement with China is more pragmatic and well intentioned than the way it is portrayed inside China or by some realists and neoconservatives internationally who lean toward an interpretation of the United States–China relationship as a zero-sum game.[82] Such views stoke the fires of mistrust and competition that do not bode well in an increasingly inter-connected world. However friendly and nonthreatening the United States' pivot toward Asia, its mere pronouncement and growing presence in the backyard of a rising power that carries a strong sense of postcolonial vic-timhood, at the very least, raises suspicions and cultivates a simmering rivalry for greater influence and power. Such perceptions matter greatly in the workings of the international system.[83]

America's pivot, moreover, has not happened in a vacuum. It has been encouraged by its Asian allies who perceive China's rise as a potential strategic threat in the region. While the United States and its Asian part-ners seek to keep China engaged instead of isolated and unchecked, they are preparing for what must be considered a likely scenario that small or larger conflicts will occur as China asserts its power. The growing and repeated tensions, for example, between the PRC and Japan do not signal an uneventful period in world affairs.

In fall 2015, for instance, tensions heated up between the United States and China after the United States performed naval maneuvers in the South

China Sea, very close to two artificially built Chinese islands. According to Zhang Yesui,[84] China's executive vice minister of foreign affairs, this move was "extremely irresponsible" and challenged Chinese territorial claims in the area. Clearly, the United States and its allies do not want to be caught off guard as China projects more of its power. In February 2016, moreover, twelve Pacific region nations, including the United States, signed the Trans-Pacific Partnership Agreement in New Zealand, a broad free-trade deal aiming to reduce taxation and trade barriers. China was not a signatory.[85]

However, in January 2017, after Donald Trump took office, the TPP was shelved[86] as part of the new President's campaign pledge to protect American jobs and to bolster the American economy against unfair competition. Nonetheless, President Trump has been a vocal critic of the PRC particularly during the election campaign even seemingly challenging the PRC's one China policy. To distance itself from the previous administration, the Trump White House has already signaled that the terms used to characterize US policy in Asia during the Obama administration will be discontinued. Susan Thornton, the assistant secretary of state for East Asian and Pacific Affairs confirmed that, "On the issue of pivot, rebalance, et cetera, that was a word that was used to describe the Asia policy in the last administration. I think you can probably expect that this administration will have its own formulation . . . [but] We're going to remain active and engaged in Asia."[87] Military exercises in the South China Sea, however, have continued under the Trump administration already drawing criticism from the Chinese government.[88] Moreover, President Trump has proven somewhat unpredictable and erratic in his pronouncements vis-à-vis US relations with both allies and rivals fueling distrust across the board. This does not bode well for the building of trust with China while it may also damage US relations with its allies in Asia pushing them more completely into the sphere of China's influence.

China's emergence indicates that the balance of world power has shifted, and this is a concern not only for the United States, but for Asian countries as well. Japan, in particular, is feeling the brunt of this change more acutely. Already in 2010, China overtook Japan and became the second largest world economy. Other countries as well are predicted to surpass Japan economically by 2030; among them India, South Korea, and other ASEAN members.[89] What would it mean if Japan were to find itself to be a middle-size power operating in a region where other actors are growing stronger?

Since World War II, Japan has reiterated its decision to be a force for stability in the Asia-Pacific region. As part of its acknowledgement of its Pacific War legacy and its attempt to redress historical events, Japan has maintained a policy of never again becoming a military power and has extended its support to the US presence in the region thus contributing to its stability.[90] In this way, Japan has managed to become more intrinsically involved economically in the region,[91] while relying for security on its alliance with the United States. In the years following the Pacific War, Japan went through four decades of remarkable economic growth, without linking its newfound wealth to a stronger political role in the Asia-Pacific region.

After the end of the Cold War, Japan was in a unique position to take over the mantle of the most powerful nation in Asia.[92] The fact that it had all the prerequisites for such a leadership role worried the United States, which had supplied the security umbrella under which Japan could perform its miraculous economic recovery from the ashes of Hiroshima and Nagasaki.[93] Japan actively cultivated relations within the region, and used its soft power to help build bonds, but, in the end, never emerged as its bona fide leader because it was not able to truly gain trust. The memories of anti-Japanese feelings dating from the Pacific War were still too fresh.[94]

Today, Japanese political and academic elites along with its public are increasingly worried about becoming marginalized as a result of their own economic stagnation and China's rapid rise. Japan believes in keeping China engaged in various regional and international fora and it strives to position itself in the role of balancer against possible Chinese domination of the region.[95] The pronounced shift in power relations and the rising tensions in the East China Sea have left Japan feeling increasingly vulnerable as China flexes its muscles in the area. Over the past few years, Japan and China have repeatedly butted heads with respect to the East China Sea in a series of incidents that included the rare earths embargo. These events underscore real security concerns and what has been described by Kazuko Mori, a China specialist at Waseda University, in Tokyo, as a "huge contradiction," with "politics and the economics moving in completely opposite directions."[96]

Currently, it is becoming more apparent that Japan may be re-evaluating its strong pacifist stand in order to respond to China's increasingly threatening rise to prominence. More specifically and according to a *New York Times* article, in 2012, Japan approved a "$2 million package for its military engineers to train troops in Cambodia and East Timor in disaster relief and skills like road building. Japanese warships have not only conducted joint exercises with a growing number of military forces in the Pacific and Asia, but they have also begun making regular port visits to countries long fearful of a resurgence of Japan's military."[97] More tellingly, since his July 2016 victory, Japan's prime minister Shinzo Abe has begun a push to revise the constitution, and especially article 9, which currently reads as follows:

> Aspiring sincerely to an international peace based on justice and order, the Japanese people forever renounce war as a sovereign right of the nation and the threat or use of force as means of settling international disputes. In order to accomplish the aim of the preceding paragraph, land, sea, and air forces, as well as other war potential, will never be maintained. The right of belligerency of the state will not be recognized.[98]

In 2015, the Japanese Parliament voted into law an important defense-policy shift that would allow troops to fight overseas.[99] This provoked considerable reaction, as it was seen as a first step in officially changing Japan's pacifist stance after World War II.[100] Abe's intentions and views about the constitution's origins, moreover, were made clear in a 2013 interview, "It is simply an illusion to argue that we, the Japanese, can take credit for opening a new, postwar era when we drafted it. In 1946, our constitution was made up by amateurs in the General Headquarters (GHQ) with no background in constitutional or international law in only eight days."[101]

To be sure there had consistently been forces in Japan that viewed the postwar constitution unfavorably from its adoption. Internal debate has been ongoing, as some political forces view the constitution as a document carrying a heavy American imprint. Nonetheless, this thinking had not represented the prevailing view, but rather reflected a smaller segment of popular opinion. It is this dynamic that is now rapidly changing.[102] Furthermore, it appears that the shifting regional balance in Asia is pushing Japan to reconsider and recalibrate the extent of its closeness with both the United States and China.

These developments, moreover, do not only affect China's Asian neighbors and the United States. While Europe does not share the close geographic proximity to China, nor does it define itself as a "Pacific power," its long involvement in Asia through trade and empire, its participation in the post–World War II institutional design of the world order, its historic and economic ties to China and the strength of its alliance with the United States are all reasons for which the EU today takes an active interest in China's rise.

Prior to the 1990s, EU-China relations were largely derivative of broader relations with the two post–World War II superpowers. The Chinese tried to use their relationship with Europe to gain strategic advantage vis-à-vis both the United States and the Soviet Union, while both the Western European countries and those of the Warsaw Pact were equally linked and constrained by their own ties to the United States and the USSR. Since the 1990s, and especially since 1995,[103] when the EU adopted its new long-term policy document on China ushering in an era of constructive engagement with Beijing, Europe and China have been expanding their bilateral relations. In the 1995 document, Europe laid out its own strategic aims in its dealings with China, seeking cooperation for shared global regional and security interests. These covered a wide range of topics: nuclear and other weapons proliferation; an international drive to contain arms sales; the encouragement of Chinese domestic political, economic, and social reform in the interests of its people; shared interests on other global issues, such as sustainable development, the protection of the environment and global resources; scientific and technological development; the information society; demographic growth; poverty alleviation; the preservation of forests; addressing the problem of illegal immigration and the control and eradication of disease; AIDS, drugs, and crime; global economic stability (given China's size and its impact as it took its place in the world system of economic rules and policies and the WTO); and issues of competitiveness.[104] Europe also aimed at raising its profile in China.

At the heart of the cooperation was Europe's and China's mutual search for economic security which led to a pronounced increase of bilateral trade. Trade, in turn, was at the foundation of the building of political relations. China is now the EU's second trading partner behind the United States, and the EU is China's biggest trading partner.[105] In fact, the two powers currently trade at the rate of over €1 billion a day. The EU and China have combined state-to-state relations with the establishment of a number of programs ranging from development aid, sectorial dialogues, various technical-assistance programs, and working groups. Europe is using this multipronged approach in order to help integrate China into the international system so that the PRC adopts a set of standards and norms that are in line with Europe's view of itself as a power intent on spreading norms and principles internationally.[106]

Europe's "China strategic plan" derives from its own historical experience following World War II, which first led its members to find an institutionalized way of dealing with their own differences, sharing resources, and collaborating to build a common future. This need was at the basis of the formation of the Union[107] itself. Europe has also sought to control the trajectories of globalization by formulating new overarching rules for the world economy and by fostering international organizations, the operations of which it actively helped design, such as the International Monetary Fund (IMF), the Organization for Economic Co-operation and Development (OECD), and the WTO. Europe is striving to use both its advantage of size as a union[108] and the influence of its individual states to build cooperative relations and to maximize its economic and strategic interests globally.

Increasingly, the EU is promoting a policy of interregionalism in order to strengthen its self-identity as a strong influential global actor.[109]

These priorities are described in the official report published by the EU Commission describing EU foreign policy objectives:

> At nearly 500 million, the population of the European Union is the third largest in the world after China and India. Its sheer size and its impact in commercial, economic and financial terms make the EU a globally important power. It accounts for the greatest share of world trade and generates one quarter of global wealth.
>
> With size and economic power come responsibilities. The Union is the biggest provider of financial assistance and advice to poorer countries. Faced with today's complex and fragile world order, the EU is increasingly involved in conflict prevention, peace-keeping and anti-terrorism activities. It supports reconstruction efforts in Iraq and Afghanistan. The Union has taken the lead in dealing with the problem of global warming and the emission of greenhouse gases.[110]

Europe is not hiding its motivations and it explicitly defends its world-view. It is seeking, moreover, to make clear that though it does not shy away from its own values, it does not seek to impose its own system on others. In describing its motivations, the aforementioned policy document released by the EU Commission, specifies that "the EU acts out of enlightened self- interest just as much as global solidarity. On an increasingly interconnected planet, supporting economic development and political stability in the wider world is an investment in one's future."[111] In this way, Europe's primary focus is in maintaining a robust multipolar system which it views as helping to curtail both geostrategic friction and brewing rivalries for superpower status that may impact its own access to markets and raw materials and hinder cooperation in a series of rising global challenges that need to be addressed.

Correspondingly, although China may not have contributed to the design of the current world order, its rise clearly impacts how the country is perceived and opens its motives to interpretation. Post-imperial modern China came into being through upheaval, war, and revolution following the abdication of the last emperor, in 1912. The bitter civil war that ended with the defeat of the Kuomintang forces of Chiang Kai-shek ushered in the People's Republic of China in 1949, which was effectively unable to play but a small part during the Cold War of fierce ideological conflict and power politics. During this time of struggle between liberal democracy and communism, China offered limited support to radical groups across the globe, principally as a way to sell arms and earn hard currency. When in the 1970s it turned its interest from Maoist ideology to domestic economic reform, it looked inward to focus its energies on growth and prosperity.[112] After Deng Xiaoping emerged the leader of the PRC, he sought to implement new economic policies and strategies.[113] There was room for much change, first and foremost in agriculture and the economy. He dismantled the commune system and abolished collectivization, which had failed miserably. He turned his attention to obsolete state factories that were reinvigorated through reforms in management. Under his leadership, Special Economic Zones (SEZ) were established; foreign scholars were welcomed on visits to China, and relations with the United States and Japan were energetically cultivated. Deng Xiaoping visited the United States, in 1979, and Japan twice, in 1978 and 1979, a first for the PRC leadership.[114]

As China grew stronger, its attention turned to the periphery once more, and it sought opportunities to build a network of friends and trading partners. Before Deng Xiaoping the PRC's foreign policy was shaped by ideological concerns and the sales of arms. Today, China's foreign policy centers on business as the driving force of its international agenda, coupled with the push to achieve a final reunification with Taiwan. So, too, China has aimed to become an actor in the world community by participating in important international organizations. According to Kurt Campbell, director of International Security Center for Strategic and International Studies,

> Beijing's policies are now functionally organized. It has market policies, energy access policies, Taiwan policies and policies designed to promote multipolarity. The pursuit of these objectives . . . is

increasingly coordinated through regional initiatives, which provide, opportunities to introduce region-specific policies to often
overlooked areas.[115]

Though China's economic growth has changed its status from developing country to international powerhouse, it still "presents itself" as the
"world's largest developing country," and continues to cultivate its relations with others in the developing world and in the region. According
to an article that appeared in the Pew Global Attitudes Project in 2007,
Andrew Kohut, the president of the Pew Research Center, discussed how
the world was increasingly developing a better view of China than of the
United States.[116] Even its neighboring countries were polling favorably
toward China in 2007. Malaysia polled at 83% favorable, Pakistan (79%),
Bangladesh (74%), Indonesia (65%), and the same went for most African
countries. The Ivory Coast and Mali gave it 92%, and favorability ratings
ranged between 67% and 81% in countries like Kenya, Senegal, Ghana,
Nigeria, Tanzania, and Ethiopia.[117] In a newer 2014 poll these ratings
remained consistent for such countries as Pakistan (78%), Bangladesh
(77%), and Indonesia (66%), and China sustained levels of favorability in
most of the previous polled African countries.[118]

Similarly, China's aim has been to promote a "harmonious society at
home," which means that it must produce results for its people. In the
aftermath of the Cultural Revolution, the PRC's one-party rule faced
irreparable ideological damage that threatened its legitimacy and ability to
govern.[119] Today, the strength or fragility of the Communist Party system
relies on its ability to continue ameliorating daily life and increasing the
per capita income of Chinese citizens.[120]

For China, a stable international environment will ensure that it is
not distracted from focusing on its domestic agenda. It even saw this
possibility—as expressed in 2002—as a "strategic opportunity" for China
to pursue its internal transformation unhindered by external issues.[121]
External stability was vital, but the next step was the quest to secure sufficient resources on a global scale to allow the government to fulfill its
domestic agenda. This search for energy led the PRC to develop a foreign

and energy policy[122] that according to some analysts has already begun "changing geopolitics around the world."[123]

Lacking sufficient indigenous energy supplies and desiring to continue fueling growth, China has been seeking to reach bilateral agreements with key energy-producing and resource-rich nations. Thus, China is actively involved in Africa, the Middle East, South America, Russia, and Central and South Asia. Oil has been a source of particular concern especially given China's growing energy needs, fluctuating oil prices, and a need for the military to plan how best to protect the sea routes across which the oil is transported to reach China.[124]

With singular focus and centralized planning, "energy security" became a stated priority in China's tenth five-year plan (2001–2005). In 2003, furthermore, Chinese authorities released a paper, entitled "Twenty First Century Oil Strategy," and allocated $100 billion to the development of a future strategic oil system and securing new energy resources abroad.[125] According to Joseph Y. S. Cheng, China's leadership, as early as 1992, had proposed that "a development strategy would have to 'fully exploit domestic and foreign resources and markets.' Since then, the three major oil companies of China, China National Petroleum Corporation, China National Offshore Oil Corporation and China Petroleum and Chemical Corporation (Sinopec Corp.), have begun their overseas acquisition programs . . . This strategy is in line with its general strategy of stepping up foreign investment to acquire resources, technology, and markets to enhance the competitiveness of its major enterprises."[126] By pursuing joint investment and technological cooperation, China is attempting to secure a strong footing in the energy market and to strengthen its ties with its partners.

China's international agenda has strong internal public support because it combines the country's return to global status and helps secure continued growth rates and increased prosperity across the board. The political narrative of success and rebirth after decades of "humiliating" defeats is embraced by the public in China and cultivated by the party. Despite all the successes achieved in the last few decades, however, the Communist Party still faces problems internally because of restricted freedoms, a growing economic disparity, allegations of party corruption, pollution, and environmental degradation. Thus, an agenda that contributes to the rise of China in the international arena helps offset some of these domestic issues.

As China's clout grows, it has focused on maintaining working relations with other large economies. It has not wanted the distraction of strained relations with other powers, so that it can continue implementing its centralized plans and objectives. Therein lies the logic of much of its strategic regional, political, and economic outreach.[127] Nonetheless, things may have already begun to change. Events have taken place suggesting that the appearance of calm might just as easily be followed by a strong storm. Recurring tensions increasingly inflect the nature of the Sino-Japanese relationship. The 2010 incident that sparked the rare earths embargo, for instance, was followed by another incident in late August of 2012 during which Japanese activists landed briefly on the islands.[128] A month earlier, activists from both China and Japan had taken turns landing on the islands—named Senkaku by Japan and Diaoyu by China. A Japanese coast guard boat claimed to have spotted a Chinese fisheries patrol boat in the area. Approximately a thousand Chinese fishing boats were reported to be moving in as well. The islands may be uninhabited, but they are thought to be resource rich[129] and are today controlled by Japan.[130] This situation worsened considerably after the Japanese state agreed to buy three of the islands from private Japanese owners, in the beginning of September 2012, for $26 million dollars.

China hastened to denounce the move as illegal and invalid. According to a news report published in the *Guardian*, Hong Lei, a Chinese foreign ministry spokesman stated, "For them to nationalize the Diaoyu islands seriously violates China's sovereignty and hurts the Chinese people's feelings. I stress again that any of their unilateral acts with the Diaoyu islands are illegal and invalid. China's determination will not change in terms of safeguarding its territory. China is observing the situation and will take necessary measures to defend its sovereignty."[131] Protesters in China attacked stores stocked with well-known Japanese goods. Protesters carried banners reading, "Never forget the national humiliation," and "Protect

China's inseparable territory," and "Let's kill all Japanese," and "Nuclear extermination for wild Japanese dogs."[132] The protests coincided with the anniversary of Japan's use of the Manchurian incident, in 1931, as pretext for the subsequent invasion of northeast China.

This turmoil produced an immediate economic impact when, on September 18, 2012, the Chinese stock market dropped and, according to *Bloomberg Businessweek*, there were concerns that "escalating tensions with Japan over a territorial dispute will hurt trade and deepen an economic slowdown." It is interesting to see what kinds of companies' stock prices were affected by the turmoil. *Bloomberg Businessweek* reported that "Chengdu Galaxy Magnets Co., which derived two-thirds of its revenue from Japan, dropped to the lowest [price] this month (September 2012)."[133] Chengdu Galaxy Magnets is one of the three largest Bonded NdFeB (MQ1) magnet manufacturers in the world.[134] The manufacture of these plastic bonded magnets relies crucially on neodymium, a rare earth. "Zijin Mining Group Co. and Jiangxi Copper Co. led a group of material producers to the biggest slump among 10 industry groups . . . The Shanghai Composite Index (SHCOMP) slumped 0.9 percent to 2,059.54 at the close."[135] The contagion of this diplomatic incident was so rapid that news reports claimed it puts to risk "a trade relationship that's tripled in the past decade to more than $340 billion."[136] Panasonic Corp. and Canon Inc. decided that under the circumstances it would be advisable to close some plants in China while protests went on, and the China Automobile Dealers Association said that "the protests will hurt sellers of Toyota, Honda Motor Co. and Nissan Motor Co. cars in China more than Japan's March 2011 earthquake."[137]

Moreover, this is not the only crisis brewing in the region. In disputed waters in the South China Sea, China has nearly completed land reclamation on certain islands and reefs. According to a statement released by China's foreign ministry, facilities would be built on these artificial "islands" for defense and other purposes, such as humanitarian and environmental functions that have yet to be clarified.[138] Tensions further increased after China's reaction to the July 2016 ruling of the Permanent Court of Arbitration in The Hague according to which "there was no legal basis for China to claim historic rights to resources within the sea areas falling within the 'nine-dash line' " (referring to the demarcation line on a 1947 map of the sea).[139]

The verdict was met by a strong Chinese reaction. "China will take all necessary measures to protect its territorial sovereignty and maritime rights and interests,"[140] the ruling Communist Party's official paper, the *People's Daily*, wrote. Furthermore, Foreign Minister Wang Yi argued that the ruling created a "dangerous situation of intensifying tension and confrontation."[141] The Chinese Ministry of Foreign Affairs went on to characterize the ruling as "null and void" and "without binding force." Moreover, in a meeting with presidents Donald Tusk of the European Council and Jean-Claude Juncker of the European Commission, Chinese president Xi Jinping made China's position on the matter very clear. According to the statement issued by China's Ministry of Foreign affairs,

> Xi Jinping emphasized that islands in the South China Sea are China's territory since ancient times. China's territorial sovereignty and maritime rights and interests in the South China Sea will not be influenced by the so-called award of the South China Sea arbitration case unilaterally initiated by the Philippines under any circumstances. China rejects any propositions and actions based on the arbitral award. China always safeguards the international rule of law, fairness and justice and sticks to the path of peaceful development. China will persist in maintaining peace and stability of the South China Sea and commit itself to peacefully resolving related disputes through negotiation and coordination with countries directly concerned on the basis of respecting historical facts and in accordance with the international law.[142]

These developments, of course, affect not only China's relations with the Philippines, which brought the case to The Hague, but also those with the other countries involved in the dispute, that is, Brunei, Malaysia, Taiwan, and Vietnam. In fact, the most recent pronouncements of Philippine President Duterte added more fuel to the fire, when he declared that he "is considering buying weapons from Russia and China while also ending joint patrols with U.S. forces in the South China Sea."[143] Earlier in the summer of 2016, China and Russia announced joint military drills in the South China Sea in September 2016. China's Defense Ministry spokesman Yang Yujun characterized them as "routine" and not in any way directed toward a third party. According to the Chinese, these exercises were meant to bolster the strategic cooperation with Russia, developing their capabilities to jointly respond to possible maritime threats that may occur. The rekindling of ties with Russia is seen as an indication of China's growing desire to block the United States' naval superiority in what constitutes a passage for over $5 trillion in annual trade.[144] Accordingly, China may have thus far opted to maintain good working relationships with its neighbors and the large economies, but these and other regional issues that have been festering for many years could quickly lead to explosive situations. It is also a warning sign that in the near future, China could become increasingly forceful and assertive of its positions.

To best see the complete picture, the wider web of policies that form the core of China's new international initiatives needs to be identified. As the PRC promotes regionalism and global multilateralism, confidence building with the states on its periphery, and a multipolar international system of power,[145] it still faces challenges and difficulties in walking a fine line. Its sheer size, economic clout, military might, and historical past are enough to keep neighboring states fearful. Moreover, its pursuit of soft-power techniques in building its relations with nations of the developing world,[146] combined with its voracious appetite for natural resources and its ability to compete in the sale of finished products, solidifies its growing global sphere of influence, but it also is beginning to raise suspicions about its strategic intentions and agenda.

There is therefore little doubt that today China is retaking its position in the small circle of powerful nations. "Though China today has an evolving definition of national interests its core objectives remain clear: to preserve a peaceful international environment; avoid encirclement or isolation; secure access to materials and markets; and promote the 'democratization' of international relations (multipolarity)."[147] This new status is under close surveillance by other major international actors, who are waiting to see evidence of a clear enough shift of priorities and interests that will lead China to abandon the difficult balancing act it has tried overall to maintain.

Already there are analysts who are seeing the beginnings of a new Chinese vision of a rejuvenated Chinese nation,[148] as it is being articulated by president Xi Jinping. On a visit to Paris in March 2014, Xi quoted Napoleon's description of China as a sleeping lion that, "when she wakes, the world will shake." According the Xi, the lion has awakened, but it is both peaceful and pleasant. There are those who think that this may not be entirely the case, and that Xi is promoting a far more expansive and muscular foreign policy, instead of following the low-profile route that Deng Xiaoping had advocated. China's prosperity is allowing it to use its economic and diplomatic power to influence the regional order. Its economic power and ever-growing trade with its neighbors, moreover, make them increasingly dependent on China for products and investment, further linking their fates and interests to China's rise and therefore marginalizing the United States as their principal political and economic partner. The fact that Japan can no longer provide the counterweight to these developments has already begun to make countries in Southeast Asia more "sensitive to PRC preferences."[149]

This kind of growing influence through China's strategic use of economic statecraft, in combination with new initiatives embarked on by President Xi, gives other powers pause. Xi has overtly revived, moreover, the concept of the Silk Road[150] connecting China to Central Asia, the Middle East, and Europe by proposing an extensive infrastructure plan for rails, pipelines, highways, and canals. He also has supported the creation of a new development bank operated by the BRICS (Brazil, Russia, India, China, and South Africa) in a bid to challenge the preeminence of the World Bank, where new powers still follow the power structure and rules of its post–World War II creators.

Furthermore, he has been the strongest backer of the creation of the Asian Infrastructure Investment Bank[151] as a new type of multilateral financial institution for investment in Asia, which was officially launched in January 2016.[152] Finally, he is tackling Asian security issues, going so far as to envision an Asia-Pacific security structure that would exclude the United States. Speaking at a conference in May 2014, President Xi underscored the point, "It is for the people of Asia to run the affairs of Asia, solve the problems of Asia, and uphold the security of Asia."

President Xi Jinping has framed his vision in ways that will not overtly intimidate China's neighbors. That is why when speaking about security in Asia, he has focused on four principles:[153] the respect of equals, a scenario of win-win cooperation, the pursuit of common security, and inclusiveness. For the moment, China seems to still have the luxury of not having to choose between strategic and economic objectives and so can pursue both "power and plenty."[154] President Xi's declared "peaceful rise," however, is not enough to put to rest the concerns in its region and beyond.[155] China's activist trade, for example, is considered by the PRC's critics to be chiefly motivated by geostrategic concerns and only secondarily by geoeconomic ones. It exemplifies China's calculated use of economic statecraft to facilitate the ultimate goal of "peaceful ascendancy."

The notion of an alternative "Chinese dream" in Central Asia and beyond[156] has been fanning geopolitical anxieties that are reflected in both scholarly and popular books and in politics in Washington. Although it has been the US position since the Nixon years that an isolated China was a more dangerous opponent than a China that engaged in global institutions, the US-China relationship has from the outset exhibited an underlying rivalry that is only strengthening as the PRC gains superpower status. Even though the United States today continues to engage China on multiple fronts, their simmering geopolitical rivalry has increasingly become the frame in which cooperation and engagement take place.

This background analysis helps to explain why the United States, the EU, and Japan reacted strongly in 2010, for both geopolitical and economic reasons. In their eyes, it exemplified China's use of a trade issue as a foreign policy instrument to advance its national interests and power. By the same token, as we shall see in chapter 4, the waning of their attention after the immediate danger subsided exemplifies their inability to formulate a more long-term strategy vis-à-vis their inordinate dependence on China for vital resources.

We are now in a position to turn to the question of rare earths in more detail and to the role that they are likely to play in these wider international developments. It will, therefore, be helpful to begin by examining why rare earths themselves have taken on such significance in the eyes of international actors.

# What Are Rare Earths?

According to the US Geological Survey, "rare earth elements (REE) form the largest chemically coherent group in the periodic table . . . and are essential for many hundreds of applications. The versatility and specificity of the REE has given them a level of technological, environmental, and economic importance considerably greater than might be expected from their relative obscurity."[1] In the development of contemporary information technology (IT) and mobility and energy technologies, a wide range of functional materials are required. Even if the quantities of rare earths necessary are small, they are nonetheless vital for the functionality of the applications.

Rare earths are elements that include the family of lanthanides (also referred to as lanthanoids)[2] plus scandium and yttrium. The atomic numbers of the lanthanides on the periodic table are 57 to 71. Scandium (atomic number 21) and yttrium (atomic number 39) are grouped with the lanthanide family because they share similar properties.[3]

Figure 2.1 depicts the periodic table and identifies rare earths.

In terms of their physical characteristics, rare earths exhibit a wide range of behaviors. Because of their atomic structures and states, they also display unique properties and thereby can be used in multiple applications.[4]

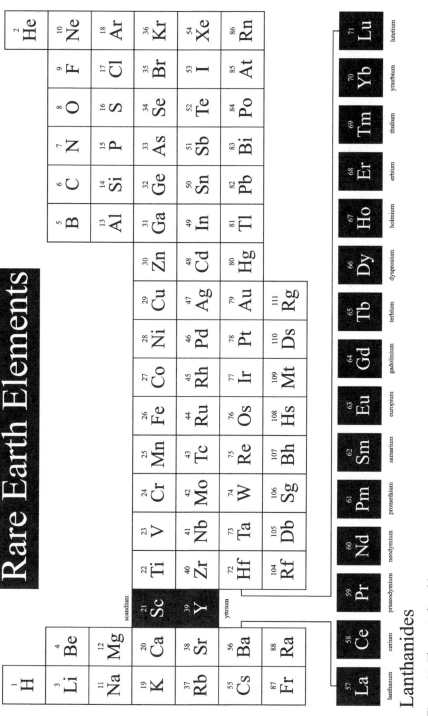

Figure 2.1 The periodic table

## RARE EARTH CATEGORIES: LIGHT AND HEAVY

Based on their electron configuration, rare earths are divided into two categories, light and heavy. The light rare-earth elements (LREEs) are lanthanum, cerium, praseodymium, neodymium, and samarium (atomic numbers 57–60 and 62). They are more abundant than heavy rare-earth elements (HREEs), which are also much more expensive. The HREEs are europium, gadolinium, terbium, dysprosium, holmium, erbium, thulium, ytterbium, lutetium (atomic numbers 63–71) plus yttrium, atomic number 39. They are used more frequently in high-tech applications.[5] Although it is the lightest of the rare earths, scandium is not typically classified under either category,[6] nor is promethium (atomic number 61) because of its radioactivity. The value of a potential mining project is, furthermore, assessed by the amount of HREE deposits that it contains, taking into consideration that they are usually more challenging to process than are LREE-rich materials. Rare earths are often separated and sold in their oxide form, designated as either light rare-earth oxides (LREO) or heavy rare-earth oxides (HREO).

Rare earths sound rarer than they are. They can be found in low concentrations throughout the Earth's crust[7] and in higher concentrations in numerous mineral deposits. In the exploratory phase, REE concentrations are reported in "parts per million (ppm)" values, and the term "total rare-earth elements or oxides" (TREEs or TREOs) indicates the sum total of rare earths in the particular deposit.

Rare-earth elements occur in most massive rock formations. What makes them rare is the low concentration in which they are discovered, ranging from ten to a few hundred parts per million by weight. The real challenge, therefore, is to find them in sufficient concentrations to make it economically feasible to mine and process them.[8]

While rare-earth elements are contained in different mineral deposits, they are most abundant in the minerals bastnaesite and monazite. Light rare earths are typically found in bastnaesite, which also contains a smaller amount of the heavier elements; in monazite the fraction of the heavy rare earths is two to three times larger than it is in bastnaesite, and the remainder comprises the light elements.[9] The mineral xenotime is third in importance after bastnaesite and monazite and contains the highest ratio of heavy rare earths. Xenotime is related to monazite and can be found in the same areas, and together they represent a continuum of mineral formation. Xenotime is a rare-earth phosphate mineral, and yttrium orthophosphate ($YPO_4$) is its main component. Xenotime is an accessory mineral in alkalic to granitic rocks, well-developed in associated pegmatite rock; in gneiss and Alpine veins; a common detrital mineral in placers.[10] Finally, the following minerals are also known to contain rare earths: apatite, cheralite, eudialyte, loparite, phosphorites, rare-earth-bearing (ion absorption) clays, secondary monazite, and spent uranium solutions.

Table 2.1 represents rare-earth mine production and Table 2.2 represents the estimated reserves for China, Australia, and the United States.

Interestingly enough, the US Geological Survey revised the rare-earth reserves in the United States to include only those "compliant with recognized standards." As a result, there was a large reduction of the reserves reported for the United States in the 2014 and 2015 data compared to the reports from previous years. In the 2014 report, for example, the US reserves according to the USGS were recalculated from 13,000,000 metric tons to 1,800,000 metric tons.[11] Australian reserves are rich in monazite, as are India's in both monazite and monazite distributed in coastal placers and inland placers. The radioactivity that comes from their thorium content, however, is a handicap because of the problems arising from the difficulties of waste disposal.

The bastnaesite deposits in China and the United States constitute the largest percentage of the world's rare-earth economic resources. The mineral bastnaesite belongs to a family of three carbonate-fluoride minerals. These include cerium-rich bastnaesite, denoted usually as (Ce, La) $CO_3F$; lanthanum-rich bastnaesite denoted as (La, Ce)$CO_3F$; and yttrium-rich bastnaesite denoted as (Y, Ce)$CO_3F$. The Swedish chemist Wilhelm Hisinger was the first to describe bastnaesite in 1838. He named it after the Bastnäs Mine in Sweden, where the ore was discovered. Bastnaesite occurs in alkali granite and syenite and in associated pegmatites. It can

*Table 2.1.* Estimated Rare Earth Mine Production (MT): China, US and Australia, 2006–2016[a]

| | 2006 | 2007 | 2008 | 2009 | 2010 | 2011 | 2012 | 2013 | 2014 | 2015 | 2016 |
|---|---|---|---|---|---|---|---|---|---|---|---|
| China | 119,000 | 120,000 | 120,000 | 129,000 | 130,000 | 105,000 | 100,000 | 95,000 | 105,000 | 105,000 | 105,000 |
| US | – | – | – | – | – | – | 800 | 5,500 | 5,400 | 4,100 | – |
| Australia | – | – | – | – | – | 2,200 | 3,200 | 2,000 | 8,000 | 10,000 | 14,000 |

[a]SOURCE: USGS Rare Earths Statistics and Information (2006–2016)

*Table* 2.2. Rare Earth Reserves, (MT): China, US and Australia, 2006–2016[b]

|  | 2006 | 2007 | 2008 | 2009 | 2010 | 2011 | 2012 | 2013 | 2014 | 2015 | 2016 |
|---|---|---|---|---|---|---|---|---|---|---|---|
| China | 27,000,000 | 27,000,000 | 27,000,000 | 36,000,000 | 55,000,000 | 55,000,000 | 55,000,000 | 55,000,000 | 55,000,000 | 55,000,000 | 44,000,000 |
| US | 13,000,000 | 13,000,000 | 13,000,000 | 13,000,000 | 13,000,000 | 13,000,000 | 13,000,000 | 13,000,000 | 1,800,000 | 1,800,000 | 1,400,000 |
| Australia | 5,200,000 | 5,200,000 | 5,200,000 | 5,400,000 | 1,600,000 | 1,600,000 | 1,600,000 | 2,100,000 | 3,200,000 | 3,200,000 | 3,400,000 |

[b]SOURCE: USGS Rare Earths Statistics and Information (2006–2016)

also be found in carbonatites and in associated fenites and other meta-
somatites. Scientists Jöns Jakob Berzelius, Wilhelm Hisinger, Carl Gustav
Mosander, and others studied the ore from the Bastnäs Mine, which led
to the discovery of new minerals and chemical elements, such as cerium
and lanthanum.

Monazite is a phosphate mineral containing rare earths. It usually
occurs in small isolated crystals, deriving its name from the Ancient
Greek μονάζειν (to be solitary).

The composition of monazite ore varies. Examples include

- monazite-Ce (Ce, La, Pr, Nd, Th, Y)$PO_4$;
- monazite-La (La, Ce, Nd, Pr)$PO_4$;
- monazite-Nd (Nd, La, Ce, Pr)$PO_4$; and
- monazite-Pr (Pr, Nd, Ce, La)$PO_4$.

Monazite ore is an important source of thorium, lanthanum, and cerium
and is found in granites, pegmatites, and carbonatites. The deposits in
India are especially rich in monazite containing large amounts of thorium
and are particularly radioactive. Madagascar and South Africa also have
large deposits of monazite sands.

The scientist Carl Auer von Welsbach was the first to notice monazite in
Brazilian sand in the 1880s. Von Welsbach became renowned for his work
on rare earths. He discovered a method of extracting thorium for incandescent mantles.[12] Monazite sand was mined in North Carolina, in addition to Brazil and, later, in Southern India when new deposits were found.
Until before the Second World War, the Brazilian and Indian mines dominated in the production of monazite. Then the mining activity moved to
South Africa and Bolivia. Large deposits of monazite are also found in
Australia.

Monazite was the leading industry choice for the production of commercial lanthanides. This changed in the 1960s, when monazite gave way
to bastnaesite because bastnaesite is lower in thorium content. Interest in
thorium as a nuclear fuel waned, and concern increased over how to safely
dispose of it after the rare earths were extracted from monazite. Today,

because rare earths continue to be extracted from monazite mines, there is an increase in R & D to find ways to make use of leftover thorium in nuclear reactors.

The monazite deposits located in Australia, Brazil, China, India, Malaysia, South Africa, Sri Lanka, Thailand, and the United States account for the second largest source of rare earths, after bastnaesite. The present worldwide estimate of monazite is roughly 12 million tons. Almost two-thirds of this occurs in deposits of heavy mineral sands on the south and east coasts of India. However, these ore deposits contain large amounts of thorium and are therefore not as attractive given the issues of radioactivity that have already been outlined and will be discussed next.

Thorium is a radioactive metal. It is more abundant in nature than uranium. Thorium is most commonly found in monazite, which can contain up to 12 percent thorium phosphate, with 6 to 7 percent on average. Finding ways to use thorium as a new primary energy source has been an ongoing challenge. The obstacle remains the difficulty in extracting its latent energy value in a cost-effective manner. Because only a small portion of the thorium extracted from monazites is utilized, most of the rest is discarded as waste. This continues to create an excessive world oversupply of thorium compounds and residues.[13] There are currently two options for dealing with excess thorium: it is either disposed of as radioactive waste, or it is stored until it can potentially be used as nuclear fuel or in the event of new applications that may require its use. This poses a significant challenge to companies that mine thorium-rich monazite ore.[14] The separation, treatment, and disposal of radioactive materials is not only an environmental hazard, but also adds considerable costs to the production cycle. Environmental regulations and the need to acquire storage and waste-disposal space explain companies' considerable investment in R & D to find commercial uses for the thorium.

Excess thorium, for example, poses a continuous problem for the Lynas Corporation in Australia. Monazite deposits there (as in India) also contain high thorium levels. In February 2012, the Atomic Energy Licensing Board granted Lynas a two-year license to begin operating its first rare earths refinery, in Malaysia. The decision sparked protests over the potential environmental hazards that could be caused by a leak of radio-active waste. In response, the board appointed an independent panel to scrutinize the plant's safety standards and asked Lynas to submit a plan for a permanent disposal facility within ten months. It also asked the company to make a $50 million financial guarantee to the Malaysian government, or its license could be revoked or suspended.[15] The two-year license was renewed in 2014 and again in 2016 until September of 2019. According to Lynas, the renewal came after a rigorous review process by the Malaysia Atomic Energy Licensing Board (AELB) and by other independent regulatory bodies in Malaysia. These developments have not altered the position of environmentalists and action groups, who continue to worry about the safety of the plant,[16] because the process of recovering thorium from monazite deposits involves leaching using sodium hydroxide at a temperature of 140 degrees Celsius followed by another more complex process to precipitate pure $ThO_2$.[17]

## RARE-EARTH PRODUCTION

Until 1948, the vast majority of rare earths came from India and Brazil. During the 1950s, South Africa took the lead.[18] After its discovery in 1949 and the beginning of its operation in 1953, the Mountain Pass rare-earth mine located in California increasingly came to lead in production until the 1980s. After mining ceased in the California mine, Molycorp—owner of the mine—sold concentrates and refined products to the market, from stocks that had been mined in the past.[19] This meant that although the United States was once largely self-sufficient in these critical materials, it became and remains dependent upon imports. The report by the US Geological Survey, is telling. According to their data, the United States, in 1999 and 2000, imported more than 90 percent of REE from China for use by American industries.[20] The United States continued to be a net importer of rare earths in 2015.[21] As for the other sources of rare earths globally, while both India and South Africa continue to produce rare earths, the amounts of their deposits in no way offset China's near monopoly.[22]

Rare earth elements (Figure 2.2) share very similar chemical properties making the separation process difficult and expensive.

The first attempts at separation began between 1839 and 1841, with the work of Gustav Mosander of the Karolinska Institute, in Stockholm. The efforts continued from 1891 to 1940, and by then, many rare-earth alloys and metals were being prepared and put to commercial use in a number of applications. From 1940–1960, moreover, effective ways of separation and development of alloys were discovered. In the 1960s, there was significant progress made in producing purer rare earths on a larger scale, which allowed scientists to identify new properties and find new uses for commercial applications.[23]

"The interesting thing about the rare earths industry is that every rare-earth deposit is different geochemically," Dr. Stephen Ward, the managing director and chief executive officer of Arafura Resources, said, speaking at a meeting in the fall of 2011. "This means unique chemistry must be developed for every resource to separate the rare earths, which is extremely time consuming and costly, resulting in a very capital intense industry."[24]

Once the elements are successfully separated, then each one, because of its unique physical properties, becomes critical material for high-tech applications in renewables, defense, or the electronics industry. This fact in itself gives the Chinese an additional competitive advantage in the separation of these elements because the Chinese have, over time, been able to master the metallurgy of their primary rare-earth deposits. What makes these elements especially valuable is that they demonstrate an extensive range of electronic, magnetic, optical, and catalytic properties.[25] REEs are considered "enablers" because they are used in the manufacturing of alloys and compounds that are then utilized in complex engineered systems.

The mining of rare earths is a complex process, however, and with the potential to cause environmental contamination requiring remedial actions over the long term. Large mines may cover areas of more than a thousand acres, and thus generate large quantities of solid wastes. As described in a 2012 Environmental Protection Agency report on rare-earth production and processing: "The primary steps involved in processing rare-earth oxides are separation and extraction from the host material

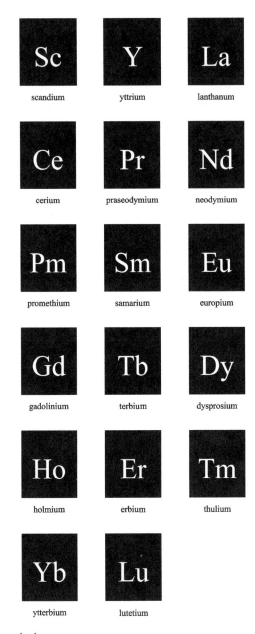

**Figure 2.2** Rare earth elements

in acidic or alkaline solutions, separation of the REOs using solvent extraction or ion exchange, and reduction of the individual REOS into pure metals."[26]

Bastnaesite mining, for example, begins by using traditional mining procedures. Then, bastnaesite must be extracted from the bulk of the ore. The remaining ore after the bastnaesite is extracted is comprised of numerous other minerals that usually have little value. The overall process of beneficiation of bastnaesite includes crushing/grinding and separation using the flotation method. More specifically, before the bastnaesite can be extracted, the ore first needs to be crushed and put through a grinding mill. At that stage, when the ore becomes as fine as sand or silt, the various minerals are separated from each other. The sand or silt is further processed to continue separating it from other nonvaluable minerals. To achieve this level of separation, the mixture must be put through a floatation process. At the Molycorp Mountain Pass Mine, for instance, even before the floatation process began, "The ore passes through six different conditioning treatments in which steam, soda ash, sodium fluosilicate, sodium lignosulfonate and steam-distilled tall oil are added to aid the separation of the unwanted materials."[27] During the floatation process, moreover, an agent needs to be used. Air bubbles are thus created that the bastnaesite will stick to and then float to the top of the tank, creating a froth-like mixture on the surface of the liquid that is then scraped off.[28]

The process does not end there. The bastnaesite itself must be separated into the different rare earths that it contains, requiring the use of acid and a number of solvent-extraction steps. Because, each and every rare-earth element has its own unique extraction and chemical processes, the rare-earth elements may need to be reprocessed until they have achieved the targeted purity.[29] Once separated, the elements are in oxide form. They can then be dried, stored, and shipped for further processing into metals. For use in such applications as the neodymium-iron-boron magnets, they are further processed into alloys. They can then be used in hundreds of high-tech applications. It takes about ten days from mining to produce the rare-earth oxides, and the process can be environmentally hazardous if not properly controlled.[30]

The process is different for monazite. Because of their high-density, monazite minerals can be released by the weathering of pegmatites and will then concentrate in alluvial sands. The geological term for these sand deposits is "placer deposits"; they are an accumulation of valuable minerals formed by gravity separation during processes of sedimentation. These deposits are often sands, and apart from containing rare earths, zircon, and ilmenite, other heavy minerals that are of significant commercial interest are also found in them. The way to produce a nearly pure concentrate of monazite is through the use of gravity, magnetic, and electrostatic separation.[31]

The two most common techniques used in monazite processing are (1) sulfuric acid digestion and (2) alkali digestion.

The sulfuric acid digestion process has been used extensively in Europe, Australia, and the United States.[32] To separate the thorium from the lanthanide in monazite, the ore is heated for several hours mixed with concentrated sulfuric acid to temperatures ranging from 120 to 150 degrees Celsius.[33]

There are variations on this basic process, and the one chosen depends largely on the ratio of acid to ore. The extent of the heating, and the degree to which water is added during leaching also determines the preferred variation. The particular methods of using acid for the separation of rare earths not only led to substantial acid waste, but also contributed to the loss of the phosphate content of the ore.

The alkali digestion process for monazite is used on a large commercial scale in Brazil and India.[34] It uses a hot sodium hydroxide solution (73%) heated to about 140 degrees Celsius. What makes this process different from the sulfuric acid process is that it permits the valuable phosphate content of the ore to be recovered, in the form of crystalline trisodium phosphate. Hydrochloric acid can then be used on the resulting lanthanide/thorium–hydroxide mixture that results in a solution of lanthanide chlorides along with a sludge of less-basic thorium hydroxide, which is essentially insoluble.

The cost of mineral extraction has increased over the years because of the lower grade ores being used as well as higher capital costs. These factors will eventually impact China's production costs as well especially if environmental and social costs are taken into account. In the past, during the

1980s and 1990s in particular, China drove down prices to help strengthen its monopoly by putting the competition out of business, but this may not be as easy for it to do again because of its own internal demand for the elements. Nonetheless, emerging economies worldwide and a continued push toward developing high-tech applications and renewables ensure that the demand for rare earths will continue to grow. Although prices have shown high volatility in the past few years following the rare earth crisis, and have taken a downward turn, things could quickly reverse as many of the long-awaited new mining and processing projects fail to go online as previously anticipated.

In the previous section it has been made evident that because the elements occur together, they need to be individually separated. The time consuming chemical baths that have been described earlier indicate that each separation venture constitutes a particularly expensive procedure that does not currently exist in a large enough scale outside China.[35] According to the industry, a separation unit costs around $500 million to build. Investing in this kind of a plant comes with a very high price tag not only because of the costly separation equipment necessary but also the environmental controls required to operate it. In addition to the high financial costs, permits take a long time to obtain, making it an even less attractive investment in today's financial climate. These financial realities attest to why it is not a simple task to break China's stranglehold on the rare earth supply chain.[36]

## RARE-EARTH APPLICATIONS

Rare earths are used in a wide range of applications that fall into two main categories. They operate as enablers in engineered or other materials and are used as components in engineered products.

Examples of REEs that are used as enablers include

- Fluid cracking catalysts (FCCs):[37] Used by petroleum refineries to convert heavy crude oil into gasoline and other important products.[38]

- Automotive emission control: Catalytic converters, catalytic fuel additives, and catalytic diesel particulate filters are used in cars and trucks to reduce emissions resulting from the engine's combustion process.[39]
- Polishing media:[40] To polish glass, mirrors, TV screens, computer monitors and the wafers that are used for the production of silicon chips.

Even in small quantities, as components in engineered products their presence enhances functionality and end-product applications.[41]

Examples include

- Permanent magnets: Not only do these generate very strong magnetic fields, but they also resist demagnetization in high operating temperatures or when exposed to other magnetic fields. Rare earths have revolutionized magnetic design, especially in electric motors and electric generators.[42]
- Energy storage: Rare earths are used in compounds to produce battery cells for energy storage.
- Phosphors: Phosphor materials emit light. This is achieved after they are exposed to electrons or ultraviolet (UV) radiation. Liquid crystal displays (LCDs) and plasma screen displays, light-emitting diodes (LEDs), and compact fluorescent lamps (CFLs) contain these kinds of materials. These are more energy efficient than older technologies and in high demand.[43]
- Glass additives: REEs are used by the glass industry to remove undesirable coloration, reduce UV penetration—a usage that is very important to protect interiors of vehicles, for example. They can also increase the refractive index of glass lenses.[44]

According to the US Department of Energy's Critical Materials Report 2011, it is valuable to examine each sector and the role rare earths play in it separately. For example, though lanthanum is used in catalytic cracking, which is an important part of the process of petroleum refining, refineries do in fact have some flexibility in adjusting their input amounts of lanthanum. Furthermore, the increases in the price of lanthanum—which is relatively more abundant than other rare earths—have been estimated to constitute no more than a penny increase on the price of gasoline, and is thus an insignificant surcharge.

Following the crisis, other industries, for example, wind-turbine and electric-vehicle companies, began looking for ways to respond to potential REE shortages. Neodymium and dysprosium are used in the magnets of wind-turbine generators and electric-vehicle motors. Manufacturers have been re-examining their product designs, seeking to redesign their systems knowing that there will be a trade-off between the efficiency that the use of rare earths provides them against the vulnerability to a supply shortage.

By way of example, in December 2010, the *New York Times* ran a story underscoring the "especially short supply of two elements, dysprosium and terbium. Tiny quantities of dysprosium can make magnets in electric motors lighter by 90 per cent, while terbium can help cut the electricity usage of lights by 80 per cent."[45] Shortages of europium, terbium, and yttrium at the height of the crisis impacted the lighting industry, at a time when new efficiency standards in lighting were being implemented.

In the defense industry, REEs are used in a wide range of systems, such as precision guided munitions, lasers, communication systems, radar systems, avionics, night-vision equipment, and satellites.[46] Furthermore, REEs will continue to be used in certain defense systems based on their life cycles. An example of one such system is the Aegis Spy-1 Radar system, which uses samarium cobalt in its magnet component. This particular system is expected to stay in use for about thirty-five years, and its magnets will need to be replaced during its lifetime.[47] Other systems that also depend on neodymium magnets, for example, are the DDG-51 Hybrid Electric Drive Ship Program and the M1A2 Abrams tank, which has a reference and navigation system that uses samarium cobalt permanent magnets.[48]

In fact, uninterrupted access to rare earths was a noted concern of the global defense industry, and particularly of the United States, which did not maintain a stockpile of these materials. In the thick of the crisis, it was the defense industry's concerns that spurred Congress to discuss the impacts of China's near monopoly. The Department of Defense (DOD) reviewed its dependency on rare earths to assess the vulnerabilities in the supply chain. They understood full well that both the commercial sector and the defense industry could be adversely impacted by a global supply chain that remained concentrated mainly in China and that produced most of the end-use applications.

Adequate mine capacity outside China, however, would constitute only part of the solution to REE shortages. Additional processing, refining, and manufacturing capacity are required to meet growing demand. This is the greater challenge because, though sustained high prices may have initially attracted investors, the market volatility has proven a disincentive to private investment. Moreover, skills and technology are also essential to build a complete supply chain outside of China, requiring time and additional investment.

Following COP21 (the UN Climate Change Conference) in Paris at the end of 2015, the world community formally recognized that the climate crisis has clearly become a global challenge that requires nations to rapidly diversify their energy mix by using renewable resources, if there is any hope for keeping the global temperature increase under 2 degrees Celsius this century.[49] Europe was among the first to take the lead and continues to support and promote a transition to a low-carbon economy. Already in 2011, the then EU Commission president, José Manuel Durão Barroso, claimed that, "Our 27[50] member states, [are bound by legislation] to reducing their emissions by 20 percent, having 20 percent of EU energy consumption to come from renewables, and increasing by 20 percent their energy efficiency by 2020."[51] Renewable energy systems rely on rare earths thus making them particularly vulnerable to supply disruptions. The following is a list of critical products and metals that are indispensable in order to decarbonize economies and stimulate economic growth:

Electric vehicle (EV) batteries: lanthanum, cerium, praseodymium, neodymium, nickel, manganese, cobalt, and lithium.

Magnets for EVs and wind turbines: neodymium, praseodymium,
   and dysprosium with samarium and cobalt as possible
   substitutes.
Phosphors in energy-efficient lighting: lanthanum, cerium,
   europium, terbium, and yttrium.
Solar cells: Thin films requiring indium, gallium, and tellurium[52]

Of these materials cobalt, nickel, manganese, lithium are not rare earths.
They are, however, on the critical metals list.

## RARE EARTH ELEMENTS: SELECTED END USES

Though the amounts of REEs used in these technologies are small, the
growth in the clean-energy sector is expected to increase world demand
for them. The clean-energy sector will face competition for these materials from other commercial tech industries (see Table 2.3) and even more
so from the defense industry, given the increased military spending pro-
jected in the Asia-Pacific region. Specifically, India's military spending was
estimated at approximately $38.35 billion for 2014–15[53] (up 11.7% from the
previous year); China's, at roughly $145 billion[54] in 2015; and Australia's at
$26.3 billion[55] Australian dollars for 2014–2015.

   Given the range of applications in which rare earths are vital inputs,
major industrial nations have taken some initial steps to address the
implications of sudden shortages in an attempt to innovate out of their
dependence on the elements. One such attempt was signaled when the
U.S. Department of Energy, in 2013, awarded a team of specialists from
around the country $120 million dollars over a period of five years to
establish an Energy Innovation Hub at Ames Laboratory, in Ames, Iowa.
Their goal is to develop solutions to the domestic shortages of rare earth
metals and other materials for US energy security.[56]

   David Danielson, assistant secretary for energy efficiency and renewable energy at the time, stated that, "The Critical Materials Institute will
bring together the best and brightest research minds from universities,

*Table 2.3.* SELECTED END USES OF RARE EARTHS

| | |
|---|---|
| Lanthanum | hybrid engines, metal alloys |
| Cerium | auto catalyst, petroleum refining, metal alloys |
| Praseodymium | magnets |
| Neodymium | auto catalyst, petroleum refining, hard drives in laptops, headphones, hybrid engines |
| Samarium | magnets |
| Europium | red color for television and computer screens |
| Gadolinium | magnets |
| Terbium | phosphors, permanent magnets |
| Dysprosium | permanent magnets, hybrid engines |
| Erbium | phosphors |
| Yttrium | red color, fluorescent lamps, ceramics, metal alloy agent |
| Holmium | glass coloring, lasers |
| Thulium | medical x-ray units |
| Lutetium | catalysts in petroleum refining |
| Ytterbium | lasers, steel alloys |

national laboratories and the private sector to find innovative technology solutions that will help us avoid a supply shortage that would threaten our clean energy industry as well as our security interests."[57] By establishing this hub, the United States effectively conceded that it had lost its edge in the rare earths industry. The hub was an attempt to address challenges across the life cycle of rare earths, seeking to improve the economics of the existing sources, but also to enable new sources. It was also set up to pursue more efficient manufacturing techniques, including reuse and recycling; accelerate material development and deployment; and to address the life cycles of new materials. The hub was designed to be involved in cutting-edge research and economic analyses of the supply chain.[58]

One central application of rare earths, is the use in the production of permanent rare-earth magnets. They are today considered the most efficient and effective magnets on the market, though it was only in the 1970s that scientists began to study them and to compare them to batteries that used ferrite materials. The continuous study of the magnetic properties of rare earths is allowing metallurgists to achieve "previously unattainable properties with even smaller, lightweight materials."[59] This is one crucial area in which China has consistently invested its efforts since the 1980s. Today, it holds the reins in all the stages of innovation and production, which in effect means that the entire supply chain for magnets is located in the PRC.

Magnetics expert Yang Luo has provided data to show the progression of China's magnetic industry over time and how it came to secure a dominant market position for its products since it turned its attention to their domestic manufacture.[60] Specifically, he demonstrates that initially, the vast majority of China's manufacture tonnage was exported. In 1998, for example 90 percent of production was exported. By 2002, which is considered a tipping point, Chinese exports equaled internal consumption. Since then, the balance has been reversed and China now exports far less than it retains for domestic consumption.

There are many reasons for this. One explanation is the rapid growth of high-tech applications such as MRIs, voice coil motors, CD pick-ups, DVD/CD Rom, mobile phones, cordless tools, electrical bikes, and electro-assisted vehicles in China. The electric bicycle has become especially popular in China, and the design of its DC motor includes NdFeB magnets.[61] Specifically, 67 percent of these neomagnets (NdFeB), which contain rare earths, are used in motors. Moreover, ongoing technological innovation and a growing Chinese middle class are pushing the use of rare earth-magnets upward. Some areas of expected growth and change are seen in the production and sales of hybrid and electric cars and trucks; in the deployment of wind turbines, especially off-shore; and the introduction of new technologies for the production of renewable energy. The market for electric bicycles continues expanding beyond China to the rest of Asia. Moreover, there is a growing demand for air-conditioning units, particularly in Asia, and a widespread use of acoustic transducers.

According to Steve Constantinides of Arnold Magnetic Technologies, though ferrite magnets remain far weaker than ones containing rare earths, they do continue to dominate in sales, representing 85 percent of permanent magnets sold. This status quo is rapidly changing, however, as technology increasingly requires lower weight and smaller size magnets to function. As a result, neomagnet (NdFeB) sales now represent over 50 percent of all permanent magnet sales on a dollar basis.[62]

By looking at the magnet industry, we can easily recognize that the benefits of having rare-earth resources available in China has allowed the PRC to bring in downstream industries from other countries. In the 1990s, for example, the United States was still the front runner in the "sintered rare earth magnet industry"; but when US mining of rare earths stopped, the magnet industry and the expertise left and moved mainly to the PRC. When the United States maintained the cutting-edge technology on these magnets, it accounted for twelve of the industry producers. Today, China holds the dominant position in an industry worth upward of $7 billion that has been projected to double by 2025.[63] Chinese companies produce over 80 percent of each of the magnet materials, and there are only a few facilities outside China that retain the capability of processing rare earths on a commercial scale.

A key technology for green energy generation is wind turbines. According to the Renewables 2016 Global Status Report,

> Wind power was the leading source of new power generating capacity in Europe and the United States in 2015, and the second largest in China. Globally, a record 63 GW [gigawatts] was added for a total of about 433 GW. Non-OECD countries were responsible for the majority of installations, led by China, and new markets emerged across Africa, Asia and Latin America . . . The offshore sector had a strong year with an estimated 3.4 GW connected to grids, mostly in Europe, for a world total exceeding 12 GW . . . Wind power is playing a major role in meeting electricity demand in an increasing number of countries, including Denmark (42% of demand in 2015), Germany (more than 60% in four states) and Uruguay (15.5%).[64]

Additionally and according to the Renewables 2017 report, an additional 55 GW of wind power capacity was added in 2016. The global total reached nearly 487 GW.

Figures 2.3 and 2.4 reflect the robust global increase in wind power capacity and that of the top ten countries.

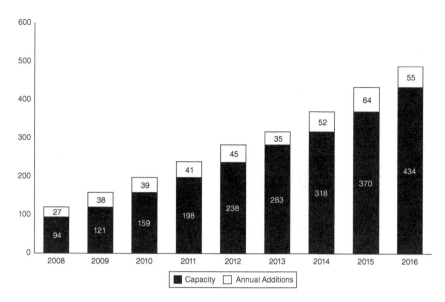

**Figure 2.3** Wind Power Global Capacity and Annual Additions (Gigawatts), 2008–2016
SOURCE: REN21 2017, Renewables 2017 Global Status Report

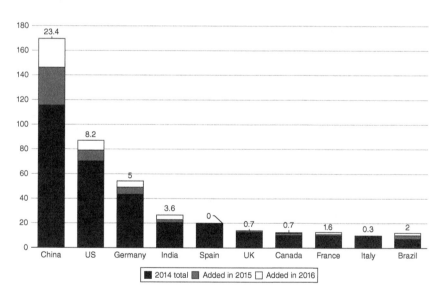

**Figure 2.4** Wind Power Capacity and Additions (Gigawatts), Top 10 Countries, 2016
SOURCE: REN21 2017, Renewables 2017 Global Status Report

According to the US Department of Energy, several design decisions have led to an increase in the use of rare earths in wind-turbine generators. One of these trends is the introduction and gradual shift to larger and more powerful turbines.[65] In 2009, 90 percent of the market was made up of wind turbines that were smaller than 2.5 megawatts (MW). The projection for 2012 altered this statistic significantly, and the share of smaller wind turbines dropped to 62 percent. This transition to larger turbines indicates that they will increasingly have to rely on the use of rare earths because these particular elements drastically reduce the size and weight of the new generators.[66]

As the climate crisis grows, so does the overall pressure to include renewables in the energy mix. In December 2015, 195 countries agreed to limit global warming to well below 2 degrees Celsius. COP 21 in Paris was thus considered an important green light for scaling up renewable energy and energy efficiency, as reflected in the Intended Nationally Determined Contributions (INDCs) of each country.

According to the Renewables 2017 Global Status Report, "Renewable power generating capacity saw its largest annual increase ever in 2016, with an estimated 161 gigawatts (GW) of capacity added. Total global capacity was up nearly 9% compared to 2015, to almost 2,017 GW at year's end. The world continued to add more renewable power capacity annually than it added (net) capacity from all fossil fuels combined. In 2016, renewables accounted for an estimated nearly 62% of net additions to global power generating capacity."[67] What is more important is that global investment remains at record levels considering the dramatic drop in fossil-fuel prices, a strong dollar, a weak European economy and continuing declines for per-unit costs for both wind and solar (photovoltaic-PV). Moreover and according to the report for 2017, "investment in renewable power and fuels has exceeded USD 200 billion per year for the past seven years. Including investments in hydropower projects larger than 50 MW, total new investment in renewable power and fuels was at least USD 264.8 billion in 2016."[68]

By the same token, permanent magnet technology plays a dominant role in the electric-vehicle market. The vast majority of hybrid electric vehicles that are produced for the mass market use rare earths in their motors. These motors, especially those of hybrid vehicles, face space constraints in their construction because of the need to fit both the gasoline engine and the electric motor into a tight space. This makes it more difficult to substitute rare earths for some other less-efficient material. The problem is also important for electric cars. Since they do not have a gasoline engine, the space constraints are not as critical; nonetheless, they, too, rely heavily on the use of rare earths.

Furthermore, technology-driven companies such as lighting manufacturers are dependent on Chinese export decisions, given that there are few other main suppliers. The supply deficits experienced, especially in 2011, led to skyrocketing prices and to increased production costs for manufacturers. In the past few years, the lighting industry has been experiencing radical changes. The new global energy standards are speeding up the phase-out process for less efficient products. Solid state LED technology has been making very quick strides. LED product entry has increased the competition and created an inflationary environment. In addition, the rise in integrated electronics is revolutionizing lighting solutions that are now being networked throughout building and living spaces. These reasons along with the significant cost increases in REEs in the last two years have had a profound effect on the lighting industry. Specifically, in less than twelve months during the crisis, the industry saw costs of some rare-earth-oxide materials used in lighting products increase from 500 percent to more than 2,000 percent before the substantial price corrections in the fourth quarter of 2011. According to a report issued by Philips (the lighting company), phosphors utilizing rare-earth oxides have been second in demand only to magnets.[69]

Rare-earth oxides of lanthanum, cerium, terbium, yttrium, and europium are used for the creation of white light from the UV radiation generated within fluorescent lamps. Phosphor is the white powder on the inside of the lamp. It is what converts the ultraviolet light into visible light. Rare-earth oxides create red, green, and blue light. It is the combination of all these lights that enables the generation of a natural and pleasant white light.

Substitution has not yet been possible. The rate of cost increase for the particular raw materials that are critical to the production of fluorescent light products and some incandescent, as well as halogen, products, has been exorbitant. Moreover, there has been a global push to begin phasing out incandescent light bulbs for general lighting uses in order to increase the use of energy-efficient new lighting alternatives, such as compact fluorescent lamps (CFLs) and LED lamps. Brazil, Venezuela, the European Union,[70] Switzerland, and Australia have also started phase-out programs. Other countries are doing the same, among them Argentina, Russia, Canada, the United States, and Malaysia.

The switch to energy-efficient lighting is considered to have a large impact not only in the reduction of carbon-dioxide emissions but also in energy savings. According to an EU announcement, "The switch to low-energy light bulbs will reduce carbon dioxide emissions by 12 million metric tons (13.2 million tons) a year and the energy saved will be the equivalent to the entire electricity consumption of Romania or the output of 10 power stations."[71] EU reports indicate that lighting may amount to a fifth of a household's electricity consumption. The least efficient lighting technologies on the market use four to five times more energy than the most efficient ones now being sold. A mere upgrade in the lamps could lead to a 10–15 percent reduction of a household's total electricity consumption.[72]

In the United States, Congress adopted energy-efficiency standards for new screw-based light bulbs in 2007. In 2012, new bulbs were required to use 25–30 percent less energy. Given that in the United States there are more than four billion screw-based sockets, expected electricity bill savings are estimated at more than $10 billion per year. This could lead to an avoidance of approximately 100 million tons of carbon dioxide per year, energy that would have been produced by thirty large power plants. If all the inefficient bulbs were replaced with the new energy-efficient kind, the savings would be equal to the amount of electricity consumed by all the homes in Texas.[73]

## THE PROBLEMS OF SUBSTITUTION AND RECYCLING

There is no doubt that the rare-earth crisis notwithstanding, there is a growing urgency for industry to adopt more sustainable practices.[74] As long ago as 1989, Frosch and Gallopoulos wrote, "The traditional model of industrial activity . . . should be transformed into a more integrated model: an industrial ecosystem. In such a system the consumption of energy and materials is optimized, waste generation is minimised, and the effluents of one process . . . serve as the raw material for another."[75] While identifying the problem, they also underscored the limitations of quickly transitioning to such an ecosystem. Nonetheless, they advocated that the concepts be widely taught to ensure that they would be incorporated in the process and in the thinking of manufacturing for the years to come. Since then, there has been growing attention to the area of industrial ecology by industry, academia, and government agencies. The expanded definition of industrial ecology, as defined by Graedel and Allenby, refers to the "means by which humanity can deliberately and rationally approach and maintain sustainability, given continued economic, cultural, and technological evolution. The concept requires that an industrial system be viewed not in isolation from its surrounding systems, but in concert with them. It is a systems view in which one seeks to optimize the total materials cycle from virgin material, to finished material, to component, to product, to obsolete product, and to ultimate disposal. Factors to be optimized include resources, energy, and capital."[76]

For these reasons alone, industry turns its attention to both substitution and recycling.

With respect to substitution for rare earths, however, alternatives are not readily available in most of the crucial applications. The substitution options to date are not as efficient or effective as those using rare earths. A good example of this is the magnets using neodymium. As yet no alternative has been developed with the strength of the magnets containing this particular rare earth. For these reasons, there is continued industrial dependence on rare earths.

Rare earths are viewed as the miracle ingredients of green-energy products, enhancing productivity and efficiency in their applications. Substitution alone would not necessarily resolve the supply issues because modern technology requires a vast range of material inputs from across the globe and there are a number of elements on the critical materials list beyond rare earths. Still, the industry invests considerable resources in finding ways to reduce rare-earth use or replace rare earths with other elements.

Just as in the case of recycling, from time to time, press reports suggest that substitution efforts are coming along. Many companies, especially in Japan, which as a high-tech manufacturer is heavily invested in the use of rare earths, are actively seeking alternatives, pushing for recycling and for becoming more resource efficient in the manufacturing process. According to the more general industrial view, however, it remains difficult to break away from rare-earth use, especially in particular applications, because they are so much more efficient and compact.

An article that ran in *Smart Planet*, in fall 2012, pointed to a few Japanese industrial efforts being undertaken to reduce reliance on Chinese rare earths. Specifically, it was reported:

> Honda Motor Corp. plans to start extracting and recycling rare earths from nickel-metal hydride batteries used in hybrid cars. In February, Panasonic Corp. introduced recycling equipment that extracts neodymium magnets from home electric appliances. Panasonic will use the neodymium—a rare earth element—in air conditioner compressors, drum washer motors, and other products . . . TDK Corp. will slash the amount of rare earth dysprosium that it uses to make magnets for cars, by developing a magnet in which it paints dysprosium onto the surface rather than mixing it into the body."[77]

Recycling is, therefore, one of the options industrial nations outside of China have been looking at as a way of addressing rare-earth shortages and in the event of future interruptions in resource accessibility. As the critical materials report published by the US Department of Energy indicates, there is much work to be done in this area because with the exception of manganese, rare earths and most other critical metals have a historical end-of-life recycling rate of less than 1 percent.[78]

Japan's efforts to recycle and recover rare earths appear to be the most pressing, and have drawn considerable attention. Although the current low rare-earth prices and technological limitations haven't allowed for these processes to widely take off to provide a viable, long-term alternative, the search for solutions continues. For Japan, in particular, given its high-tech industrial output and its proximity, dependence on, and growing competition with China, the desire to reduce its dependence appears to be the most urgent concern driving further research and innovation. Since the crisis, it has been frequently reported that companies, such as Hitachi, Honda, Japan Metal and Chemicals, and Umicore, have announced technologies for recovering rare earths. However, many of the details of these technologies are not yet widely available.

One viable area that Japan has turned to is recovery from end-of-life-vehicles. The focus is on the recovery of NiMH (nickel metal hydride) batteries and NdFeB magnets.[79] In the past, because rare earths have not only been used in small amounts but have also been dispersed through numerous vehicle components, recycling had been negligible. In 2010, less than 1 percent of REEs were being recycled from end-of-life products. Inefficient collection, technological difficulties, and a lack of incentives to scale up lab research has kept these figures low.[80] Today, because of technological innovations, there are increasingly large amounts of rare earths in use, with a particular emphasis on dysprosium and neodymium being used in vehicles. In a conventional sedan, for instance, 0.45 kilograms of rare earths are used while for a hybrid electric vehicle (HEV) with an NiMH battery the amount is approximately 4.5 kilograms.[81]

A scenario analysis by Guochang Xu, Junya Yano, and Shin-ichi Sakai in fact presents potentially encouraging findings for the recovery of rare earths from end-of-life vehicles (ELVs) by 2030,[82] although the authors recognize the complexity involved, not only in the process itself, but because of the wide range of vehicles and the different methodologies used to assess the content of rare earths in different components. There are other areas, such as electric power steering and electric air compressor units, where recovery may prove viable in the future, but at the moment results remain limited to laboratory-scale findings.

There seems to be a desire to show success in this area in order to prove that China's hold on the elements is not absolute; however, the reality remains that the recycling and recovery of rare earths from a wide range of applications is still in its infancy. A recent EU report on electronic-waste recycling continued to quote rates of less than 1 percent, pointing to the fact that for any move toward industrial-scale recycling of REEs to take place, new technologies not only need to be developed but the economics also need to be sound. Specifically, the "costs of the process, the need to achieve economies of scale and REE prices and the absence of a supply chain structure geared towards the pre-processing of WEEE (waste electrical and electronic equipment) with focus on REEs" are vital links to the chain of recycling and recovery.[83]

There are nonetheless multiple reasons to pursue recycling. It reduces the environmental impact of increased mining and the waste and toxins that are discharged and discarded. It also creates a secondary supply source. This is one of the areas of strong cooperation among the US, the EU, and Japan as they seek to address rare-earth shortages. This is not as simple an endeavor as it might seem, however. Its success is dependent on a number of factors: material availability, first and foremost; effective recycling technology; logistics; and economics. The last two are very important because the collection system of rare earths in diverse technologies is a difficult operation. The technology and infrastructure require significant capital investment and R & D. In a world market that is still in recession, raising the needed capital is not a given.

The success of recycling, if it is not centrally mandated, imposed, and funded, depends largely on the current price of rare earths. If the prices are low, then recycling will not be economically viable. The cost of collection, transportation, characterization, sorting, separation, purification, and other processing[84] all make for a high cost of recycling, in general, and of recycling particular elements. There are significant technical difficulties in separating the rare-earth elements from scrap because of the small amounts in which they are found in applications.[85] Furthermore, rare earths such as lanthanum and cerium are abundant and are usually underpriced (in fact, they are sold below the average production costs). It is the heavier rare earths that make up for the price difference for REE producers. This means that the recycling of the heavy and more expensive rare earths would make more sense than that of the light ones.

The lighting industry could be attractive for end-of-life recycling. Spent fluorescent light bulbs are currently recycled at around 30 percent to remove the mercury. Unfortunately, most of the rare-earth-containing phosphors are thrown into landfills, where most spent bulbs end up. Technological challenges aside, this could be an area where recycling could bear fruit.

Beyond substitution and traditional recycling, however, there is a push to evaluate the extent of waste materials from mining operations and from the industrial processes. These types of solutions help create a culture of resource efficiency and lead to a significant reduction of harmful waste. This is an area in which the EU[86] in particular is keen on making strides. Similarly, one particular area that could be of interest is the effort to reduce manufacturing losses in the magnet industry. Currently, it is estimated that approximately 30 percent of the magnetic material is lost during the machining process, and this certainly presents opportunities for improvement in the recovery rates.

## ENVIRONMENTAL CONCERNS OVER REE PRODUCTION

Environmental pollution has been a problem with regard to rare earths and contributed to the reduction of mining activity in industrialized nations such as the United States, where concerns associated with the mining of monazites and their radioactivity nearly wiped out REE production. The disposal of radioactive material is costly and environmentally hazardous.

China, on the other hand, did not enforce strict environmental oversight and regulation of its mining sites, and the purification process. This led to substantial environmental degradation and pollution of the areas around the mines and nearby rivers for decades. This is one reason the government claims to have clamped down on the illegal production and smuggling of REEs outside China.

According to a document published by the Chinese Society of Rare Earths, and quoted in an article on *The Cutting Edge* in November 2010: "Every ton of rare earth produced generates approximately 8.5 kilograms (18.7 lbs.) of fluorine and 13 kilograms (28.7 lbs.) of dust; and using concentrated sulfuric acid high temperature calcination techniques to produce approximately one ton of calcined rare earth ore generates 9,600 to 12,000 cubic meters (339,021 to 423,776 cubic feet) of waste gas containing dust concentrate, hydrofluoric acid, sulfur dioxide, and sulfuric acid, approximately 75 cubic meters (2,649 cubic feet) of acidic wastewater plus about one ton of radioactive waste residue (containing water)." [87] The enormity of the environmental impacts on the freshwater system cannot be understated. Without effective treatment, releasing this wastewater into the environment poses a direct threat to both the drinking water supply as well as to the water supplies used to irrigate the nearby fields.

Tailings are another dangerous source of environmental pollution. In many cases, the rare-earth tailings contain considerable amounts of radioactive thorium and therefore require special treatment. The problem of dealing with the leftover thorium is particularly complex and difficult and poses a long-term issue not only for China, but also for the Lynas Corporation which owns and runs the largest rare-earth mine in Australia. Earlier, the strict regulations necessary for the safe disposal and storage of radioactive materials, involving a number of permits and controls, were largely responsible for the initial closing of the US Molycorp mine.

Until recently, the environmental repercussions[88] of rare-earth mining had not been robustly addressed in China. According to a report based on findings by the Chinese Ministry of Industry and Information Technology, for example, the clean-up costs for polluted rare-earth mines in Ganzhou alone are estimated at $6 billion.[89] What is worrisome is that the Jiangxi province represents only 8.6 percent of the Chinese rare-earth production. Thus far, the government appears to have spent $3.8 billion yuan (about $569 million) for the restoration of mines in the area. Moreover, it appears that this exceeds the gains made by the rare-earth companies operating in the particular area.[90] Much of the damage is attributed to extensive illegal mining practices but nonetheless, it becomes clear that the environmental cost had thus far not factored into the price of rare earths. In Bayan Obo, for example, the tailing dam in operation is one of the world's largest. The radioactive toxic waste leaches into the ground, impacting the health of those living in the nearby villages. The Yellow River is only 10 kilometers away, putting at risk one of the main sources of fresh water for millions of Chinese.

The process of refining rare earths is also highly problematic because it requires the use of chemicals such as ammonium bicarbonate and oxalic acid. Numerous and severe health hazards stem from the use of these chemicals. The inhalation of ammonium bicarbonate, for example, can cause irritation to both the respiratory and gastrointestinal tracts. Oxalic acid is poisonous if swallowed and possibly fatal. It is corrosive and can damage the kidneys. Given the abysmal environmental checks and safety measures employed in the refining process, these kinds of chemicals have been contaminating the waterways of China.[91] In Jiangxi, moreover, illegal processing plants have been pumping acids directly into agricultural fields, causing contamination and ruining crop production. The extent of these environmental impacts has led to protests among the local population.[92]

The rare earths industry in China until now has been large and unruly. Part of the reason companies did not invest in the creation of a more environmentally friendly mining practice is purely economic. When rare-earth prices were low and the government pressure was not there producers opted to maintain a competitive edge in the market. In addition, the Chinese government owns the land, further deincentivizing companies from undertaking the clean-up investment. Furthermore, there was no support provided to companies for implementing environmental standards.

The Chinese government has now stepped-up measures to rectify the situation. Environmental degradation was, in fact, one of the government's main arguments for deciding to more strictly monitor and regulate the industry. China's official environmental concerns were voiced in a white paper addressing the concerns of the international community, distributed in the second half of 2012, according to which,

Outdated production processes and techniques in the mining, dressing, smelting and separating of rare earth ores have severely damaged surface vegetation, caused soil erosion, pollution, and acidification, and reduced or even eliminated food crop output. In the past, the outmoded tank leaching and heap leaching techniques were employed at ion-absorption middle and heavy rare earth mines, creating 2,000 tons of tailings for the production of every ton of REO (rare earth oxide). Although a more advanced in-situ leaching method has been widely adopted, large quantities of ammonium nitrogen, heavy metal and other pollutants are being produced, resulting in the destruction of vegetation and severe pollution of surface water, ground water and farmland. Light rare earth mines usually contain many associated metals, and large quantities of toxic and hazardous gases, wastewater with high concentration of ammonium nitrogen and radioactive residues are generated during the processes of smelting and separating. In some places, the excessive rare earth mining has resulted in landslides, clogged rivers, environmental pollution emergencies, and even major accidents and disasters, causing great damage to people's safety and health, and the ecological environment. At the same time, the restoration and improvement of the environment have also heavily burdened some rare earth production areas.[93]

Although it is ironic that these miracle elements of high-tech and green technologies have such serious environmental impacts when not processed in accordance with strict environmental regulations, they remain indispensable in modern applications that the world increasingly depends on.

The fact that China holds the reins of their production, excelling not only in their mining, but also in the metallurgy needed to successfully separate them, should give industrial nations pause. Innovating out of the use of rare earths is not a likely prospect in the decades to come. Neither is extensive recycling, especially not at the current, post-crisis prices. On the contrary, most of their important applications by all predictions will be increasingly needed in the future to power the digital world and to help us decarbonize our economies.

# Salt and Oil

*Strategic Parallels*

Rare earths are, of course, not the first resources to be considered important in the construction of our man-made environment. Other renewable and nonrenewable resources have been coveted by nations throughout history. Few, however, have managed to rise to the level of importance to be considered "strategic." Among those very few, salt and oil have been perhaps the most prominent.

Interestingly enough, however, they did not share the same trajectory and fate. Salt's critical importance, chiefly as a food preservative, dates back to antiquity and lasted until the end of the nineteenth century; however, today it is considered to be a mere condiment on our table, a substance found in abundance. It was innovated out of its strategic importance by new technologies that rendered some of its most vital uses obsolete. Oil broke onto the scene after Edwin Drake drilled the first oil well in northwestern Pennsylvania, in 1859, and Nikolaus Otto invented the first gasoline-burning engine, in 1861. Since then, over fifty thousand oil fields have been discovered around the globe (though over half of the production still comes out of the forty largest ones). Oil continues to power the modern economic miracle. Although the climate crisis is pushing the world to, at the very least, begin to diversify its energy mix with renewables and nuclear energy, oil dominates the energy choices for a global population of seven billion people.

I believe that it is helpful to examine the cases of salt and oil because it allows us to draw analogies with the present case of rare earths, which also have become indispensable to our way of life because of their centrality in a wide range of technological applications that define our era. An overview of the cases of salt and oil as strategic commodities highlights the importance that certain materials have had in shaping history, innovation, and the global economy. Furthermore, both resources draw our attention to China's historically centralized approaches to securing and controlling them, providing us with a more nuanced framework for a deeper understanding of present Chinese attitudes in formulating policy vis-à-vis rare earths.

## SALT

First, I will trace the rise and fall of salt as a strategic commodity and employ its long historical trajectory as an important comparandum for isolating key features in the case of rare earths. Both the perception of salt as a strategic and rare commodity and the deliberations of international actors in light of that perception offer a series of important historical parallels. Conflict, cooperation, use of economic statecraft to gain advantage, and technological innovation are all part of salt's long global history. Moreover, our evidence of Imperial China's methods for securing what the Chinese took to be a strategic commodity is likewise extremely suggestive since they recognized the critical value of salt and developed methods to secure, tax, and trade it, while extending its uses through technological innovation. China's particular approach, moreover, is reasonably well documented and offers an intriguing window into its tendencies toward centralized decision-making.

Ancient Imperial China developed an intricate system of taxation and methods of administering the revenues that salt brought into the imperial coffers. Its importance also spurred technological breakthroughs in the production of greater quantities more efficiently, in many instances, much earlier than in the West. Once the central Chinese authorities developed a perception of salt as a strategic resource, they wasted no time in devising methods to control it, not only through taxation, but also by administering its production and distribution. China's policy exhibited a pattern of the internal control of salt as an important strategic asset.

Other empires and major powers also recognized salt's importance, and their policies reflected this understanding. Salt became the driving force behind international commerce, especially in the fish trade in which its use as a preservative was crucial, and it became one of the focal points of great power rivalry in the New World. Often, the rivalries over access to salt led to wars, wars which themselves were often not sustainable without access to abundant salt supplies. Centralized control of salt by governments worldwide often led to high prices, which in certain instances became cause for rebellion.

Eventually, the understanding of chemistry and the geology of salt made it clear that it was not a scarce commodity at all. It was technological innovation, however, that finally deprived salt of its most strategic application—namely, preservation. Today, salt is just a simple commodity sold on the international market. Given the loss of both its strategic importance and geographic singularity, it can no longer generate income for government coffers, and its centralized control through high taxation no longer has any economic justification.

The brief catalogue that follows can perhaps serve as an adumbration of some of the possibilities underlying the present story of rare earths. It also raises questions about whether these vital elements may lose their strategic importance through new discoveries and technological innovation.

In today's world, we think of salt as little more than a condiment that we add to our food, and estimates of per capita consumption range from two-thirds of a pound to sixteen pounds per year, depending on the climate and the work that individuals are engaged in.[1] Salt, however, has a much larger historical footprint than that of a mere accouterment in the kitchen or on the dinner table. Since Neolithic times, salt was also used as the most important preservative, thus transforming it into the single most sought after commodity in human history.[2]

Many might find it hard to believe that it was only after the nineteenth century that international maneuvering and strife over the acquisition of salt came to an end. In his seminal book, *Histoire du sel*, Jean-Francois Bergier, draws a clear parallel between salt and oil in terms of their importance:

> It is only since the 19th century, and even though the consumption of salt had increased due to its new industrial uses that the international market for this product finally waned. Before, salt had sparked intense trade. It constituted an object of much speculation on the part of producers and caused anxiety of consumers who were rarely, in the long-term, assured of satisfactory supply. Salt had justified the design of both trade and political strategies. It brought wealth to some and impoverished others. In short, for scores of generations salt played the role that ours has assigned to oil. Same universal use, same scarcity in terms of geographic distribution of accessible resources. In a word, as a result, the same tensions between producer and consumer countries, the same political fiscal temptations.[3]

Salt served as currency, engendered trade routes and cities, and provoked and financed wars.[4] Salt helped to build strong empires and provoked revolutions. Some scholars have argued that the earliest roads were built to transport salt, and the oldest cities served as centers of the salt trade.[5] Salt's importance through the ages is intrinsically woven through the world's religions, traditions, history, and in trade and innovation.

Looking back in time, we see the range and versatility of its applications. The ancient Egyptians used salt for mummification. Jews and Muslims sealed their deals using salt as a symbol of the transaction's solidity. The ancient Greeks, Egyptians, and Romans used salt in sacrifices and offerings.[6]

On a more practical level, after people began cultivating the land, salt was needed to supplement their diets, because crops alone did not provide enough sodium for survival. In contrast, tribes who relied heavily on game did not have the same dietary need because they got the salt they needed from meat.[7] Salt also became important after animals became domesticated and were fed by their keepers. Throughout history, salt's potential to help raise state revenues was not lost on governments. Imperial China, the Romans, the French, the Venetians, and the Hapsburgs, to name merely a few, used taxation, regulation, and state monopoly to control this valuable asset.

In historical accounts of salt production in China, there are roughly four periods of development. The first period ranges from 4000 BCE to about the fourth century BCE, when salt was harvested from naturally occurring brine springs and from seawater. It was also found in salty lakes.[8] The second stage covers the period between the fourth and third centuries BCE and is characterized by technological developments, such as an increase in the use of boiled seawater, the building of a system of transport across long distances and the widening of the mouths of brine wells to get better access to brine. The third stage covers the period from the Han dynasty to the eleventh century; and the final period includes the beginning of the twentieth century, when production had reached a larger scale through the invention and implementation of new technologies. It is in the last period that the Chinese made strides in drilling techniques, which allowed them to reach depths of 1,000 meters.[9]

China's clear-cut strategy in dealing with salt began as early as 2000 BCE. Texts from that period have been found showing that salt was being taxed. Indeed, it became a main source of state revenue and, because of its importance, a focus of impressive technological innovation.[10] Because salt was so vital to people's livelihoods and well-being, China viewed it as the ideal resource to tax because everyone could thus contribute to the state's finances. Nonetheless, before the unification of China's warring states, there had been much debate about how and whether to tax salt.

Some historians[11] attribute the policy of government control of salt to Guan Yi Wu, a renowned strategist of the seventh century BCE. He came from the feudal state of Chi (Qi).[12] Guan Yi Wu explained to the ruling prince of the state how the creation of a salt monopoly, given statistical patterns and consumption, would bring him considerable riches. The state of Chi adopted this policy and benefited greatly from the new practices.

There are however, earlier mentions of salt as "tribute" to the imperial household, for instance, in the *Canon of History* covering the years 2400 to 619 BCE.[13]

By the time the Qin dynasty came to power and ruled over a unified China, it was decided that both salt and iron would be controlled by the state, in a price-fixing monopoly scheme. The monopoly kept prices very high. According to Mark Kurlansky, this was "the first known instance in history of a state-controlled monopoly of a vital commodity."[14] The government levied the salt tax in lieu of a poll tax.[15] Money from salt taxation was used to raise armies and to build defensive works, including the Great Wall of China, to keep out the Huns and other nomadic peoples.

The debate over whether or not to have a state monopoly over iron and salt in China reflected two competing schools of thought: Confucian versus legalist. The Confucians were mainly concerned with the "social economy"; whereas the legalists weighed in on the importance of "enriching the state."[16] In the book *Discourses on Salt and Iron*, which dates back to the first century BCE, Huan K'uan provides a revealing glimpse of the debate between the legalists and the Confucian schools with regard to salt (and iron) policy. The defense of China was the principal justification for the state's need to have a monopoly, so that it could raise the necessary funds for the undertaking. Accordingly, the Lord Grand Secretary said,

> When the Hsiung Nu rebelled against our authority and frequently raided and devastated the frontier settlements, to be constantly on the watch for them was a great strain upon the soldiery of the Middle Kingdom, but without measures of precaution being taken, those forays and depredations would never cease. The late Emperor, grieving at the long suffering of the denizens of the marches who lie in fear of capture by the barbarians, caused consequently forts and a series of signal stations to be built, where garrisons were held ready against the nomads. When the revenue for the defense of the frontier fell short, the salt and iron monopoly was established, the liquor excise and the system of equable marketing introduced; goods

were multiplied and wealth increased so as to furnish the frontier expenses.

Now our critics here, who demand that these measures be abolished, at home would have the hoard of the treasure entirely depleted and abroad would deprive the border of provision for its defense; they would expose our soldiers who defend the barriers and mount the walls to all the hunger and cold of the borderland. How else do they expect to provide for them? It is not expedient to abolish these measures![17]

To this criticism the Literati responded that

the Son of Heaven should not speak about much and little, the feudal lords should not talk about *advantage and detriment,* ministers about *gain and loss,* but they should cultivate benevolence and righteousness, to set an example to the people, and extend wide their virtuous conduct to gain the people's confidence. Then will nearby folk lovingly flock to them and distant peoples joyfully submit to their authority. Therefore, *the master conqueror does not fight; the expert warrior needs no soldiers; the truly great commander requires not to set his troops in battle array* . . . The Prince who practices benevolent administration should be matchless in the world; for him, what use is expenditure?[18]

Given these two "warring" schools of thought, monopolies were at times stopped and then resumed when the state's coffers were empty and it needed money. The monopolies at times led to popular uprisings protesting the high prices that they caused. Nonetheless, salt monopolies did endure over time as attested by the Chinese word for salt, *yan.* It is a pictograph in three sections: the lower part depicts tools; the upper left depicts an imperial official; and the upper right shows brine.[19] The government showed its direct involvement in salt from the outset.

Throughout China's long imperial history, the Chinese "gabelle" (taxation system) was generally handed down from one dynasty to the next.

The Ming dynasty was the first to reorganize the system to achieve greater efficiency and consolidate its control.[20] During the Ch'ing dynasty (1644–1911) the salt supply, being widely dispersed across the country, was under strict and elaborate bureaucratic state control. A chief goal of this control was to eliminate smuggling and thus to increase government revenue.[21]

By 779 CE, salt accounted for half the central government's intake. But salt production in China did not only generate valuable revenue.[22] It also spurred technological innovation. It would take many centuries for European nations to catch up. Lin Bing—by all accounts one the most important hydraulic engineers of all time—was governor of Shu, today known as Sichuan, during the Warring States period. Shu was a salt-producing area, and according to early records, Li Bing discovered that the pools of natural brine from which the region's salt was made originated from salt seeping up to the surface from underground. In 252 BCE, he commanded the drilling of what were the first brine wells in the world.[23]

The Chinese worked hard to learn and perfect the art of drilling to extract brine and make salt. They innovated far earlier than the Europeans and went deeper and deeper into the ground. They used long bamboo tubes that they lowered into the shafts to bring up the brine, and then they transported it to the boiling houses for processing. By the eleventh century, the Chinese had developed percussion drilling, which remained the most advanced method of drilling for another seven to eight centuries. By the end of the eleventh century, the Chinese had also discovered that a mixture of potassium nitrate (known as saltpeter) with sulfur and carbon could be lit to produce an explosion. The result was gunpowder, which was the first widespread industrial application of salt.[24]

Whereas the salt trade in Western Europe took place across national borders, China's production was mainly for domestic consumption. Overall, as Zbigniew Brzeszinski points out,

the Chinese system was self-contained and self-sustaining, based primarily on a shared ethnic identity, with relatively limited projection of central power over ethnically alien and geographically peripheral tributaries. China was quite unlike other empires, in which numerically small but hegemonically motivated peoples were able for a time to impose and maintain domination over much larger ethnically alien populations.[25]

Nonetheless, the Chinese were not the only civilization that was aware of the benefits of salt's properties and how to use it as a tool of trade. The Egyptians mummified their pharaohs and preserved their food with salt. In the north of Europe, the Celtic people had developed many industries, including salt mining and salt trade, as well. Many of their contributions were taken over by the conquering Romans.[26] The Romans did not maintain the kind of monopoly on salt that the Chinese did, but they did step in to control salt prices when they deemed it necessary. Under both the Republic and the Empire, Rome would subsidize salt prices to ensure that it was affordable to plebeians, especially when their support was needed for a military campaign.[27] Marcus Livius designed a system for taxing salt. He became known as the "salinator" because of his price scheme.[28]

Gaining access to salt was part of the Roman conception of state building, which they accomplished by developing saltworks across the empire, using salt collected from seashores, marshes, and brine springs. They also annexed those from the conquered lands of the Celtic peoples and also the Phoenicians and Carthaginians. Sixty Roman saltworks have been identified.[29] Romans focused less on inventing new extraction techniques and more on administering and distributing, thus creating a "worldwide" network of saltworks. This "global" operation changed hands as power passed from one empire to the next, from the Romans to the Byzantines to the Muslims.

By the late 1200s the Venetians, who controlled maritime trade, discovered that they stood to make more money by trading salt than by producing it themselves.[30] This turned high profits for the city, allowing it to undercut competitors' prices in the transportation of other goods, especially the valuable Indian spices from the eastern Mediterranean that they sold to Western European countries.[31] Instead of taxing salt, Venice regulated its trade to make money, and the city became a reliable supplier of salt to many countries of the world, signing exclusive contracts with a number of them. The Venetians searched for new areas of production and sought to control supplies and saltworks by buying them wherever they were available for purchase.[32] That is how Venice manipulated and controlled the market. As Venetian power grew, the city-state not only manipulated the market but sought to control territory by force, even if it meant war.[33]

In another part of the European continent, from the seventh century CE onward, the Basques became the European continent's expert whalers and, later, cod fishermen who preserved their catch in salt. Salting cod proved to be an enormous business because after the salted fish was soaked in water, it was lean and white. The profitability of this venture attracted other northern European powers. For the Scandinavians, the fish of choice to salt and sell was the plentiful "herring," preferred by the poorer classes. Herring needed to be salted within twenty-four hours of being caught, which again meant that immediate access to salt for curing was the issue that needed to be resolved.[34]

By the Middle Ages, salt had acquired a whole range of applications. It was used to cure leather, glaze pottery, clean chimneys, solder pipes, and for medicinal purposes, as well.[35] Years later, the British and French began to value salt as a strategic commodity, vital to war efforts, because of the salted cod and corned beef that were used as rations by the British Navy. In fact, by as early as the fourteenth century, salt had become valuable in war preparation because of its use for food rations.[36]

In France, dairy farms tried to find ways to preserve milk using salt. Different climates and local traditions led to the creation of at least 265 different kinds of cheese, for which France is renowned to this day. So important was salt in France that the state imposed a taxation system that continued to the twentieth century.[37] In France, the gabelle was resented because, as a poll tax, it charged the poor and the rich alike for the use of a commodity that was vital for life itself. It reinforced the rarity of salt because of the ways that the taxation affected trade. Furthermore, the gabelle, which initially began as a modest 1.66 percent sales tax on salt by the French crown, had by Louis XIV's time become a leading source of state revenue. The gabelle was not a universal uniform tax but different in different areas of France,[38] a practice that led to smuggling and unrest.

Salzburg, too, owed its riches to salt. While the city also possessed gold, copper, and silver, it was salt that gave it its extraordinary wealth. Understanding full well the importance of controlling salt, the Hapsburg Empire established a monopoly as it expanded its influence and took over salt production in central Europe.

The battles over access to salt and fish kept growing, as trade expanded and substantial profits could be made. Competition shifted to the New World, where the British managed to acquire the lion's share of codfish grounds in North America from 1713 to 1759, after beating out France.[39] The Europeans forcefully took over control of salt from the indigenous population. The Great European Powers spent the seventeenth and eighteenth centuries battling over the control of the islands of the Caribbean, interested as they were in both salt and sugar cane. In fact, salt was considered such a valued commodity that ships carrying it traveled in convoys to avoid being seized by enemy warships or pirates.[40]

The salt wars were on in the Americas.[41] The Massachusetts colony granted monopolies to those who knew how to produce salt, and the colony gave Samuel Winslow a ten-year monopoly—the first patent recorded in the Colonies—to employ his techniques for salt making. Salt was needed for salting cod, but also in the fur trade, because the locals salted the bear, beaver, moose, and otter skins for preservation purposes during transport. The typical New England family also needed access to large amounts of salt because of the way they planned their food needs for the year. They slaughtered meat in the fall and preserved it in salt, for example.

By 1775, war was afoot because the British insisted that Americans import salt from them. There was a salt shortage in the Colonies[42] and ways to get it were aggressively sought. The American War of Independence took place against the background of a salt shortage.[43] After the treaty marking the end of the war between the United States and Britain was signed in Paris, the conflict over salt continued. Americans turned to innovation to become salt sufficient. The next major technological breakthrough to affect the salt trade was the invention and use of the steamboat, after

1790, which allowed for heavy salt shipments to travel by river into the US interior. This, together with the successful completion of the Erie Canal, led to proposals for more projects connecting the major rivers across the interior.

What came next was the Civil War. By 1858, the states in the American South—namely, Virginia, Kentucky, Florida, and Texas—were producing 2,365,000 bushels of salt, while New York, Ohio, and Pennsylvania in the North produced 12,000,000 bushels.[44] Americans' salt consumption grew and began to outpace that of the Europeans. The South relied on salt imports from England and the British Caribbean; the North was more self-sufficient. But there could be no war without salt. It was necessary for the soldiers' diets and for medicinal and disinfectant purposes in treating the wounded.

In her book *Salt as a Factor in the Confederacy*, Ella Lonn includes an interesting anecdote showing how crippling it was for the South to not have access to its own salt supplies.

> An ex-confederate officer was giving a lecture in Syracuse, New York, a few years after the close of the Civil War. In the course of the day he had been driven about the city by his host and had, naturally, been given a view of the extensive salt wells and salt works for which the city was noted. In opening his lecture that evening he startled the audience with the somewhat remarkable query, "Do you know why you northerners whipped us southerners?" On the surprised ears of his audience fell the terse answer, "Because you had salt."[45]

Salt played a key role in the history of India as well. By late 1929, Gandhi was asking for independence from the British Empire. It had been an extremely difficult period because the great depression had taken a toll on the Indian economy. Gandhi announced an eleven-point program of various demands that would appeal across the India social groups. One of these demands was the abolition of the salt tax. The salt tax netted 25 million pounds for the British Empire compared to the 800 million pounds collected by the British from India overall. Salt, however, held symbolic and practical meaning for the Indian population.

Although salt was abundant across India, the manufacture of salt had long been a government monopoly, and the British made it a punishable crime to possess salt from a nongovernment source. This policy had been criticized in the past, but Gandhi used it creatively to defy the laws in order to challenge British authority in India.[46] In March 1930, the Salt March began. On April 6, Gandhi waded into the sea as a symbolic act of purification.[47] Then he passed a spot where the salt was thick and took some in his hands saying, "Our path has been chalked out for us. Let every village manufacture or fetch contraband salt."[48] The salt movement began to spread, and teachers, professors, and students, who had begun collecting their own salt, ended up in jail in growing numbers. They never resisted arrest, but they did resist the confiscation of the bags of salt they had collected. They were beaten for this. Gandhi was arrested, and protesters who had not lifted a finger against the authorities were beaten by the police forces. The salt protests spread across the Indian coastline. Eventually, on March 5, 1931, the British arrived at an agreement with Gandhi that ended the salt campaign. Among other things agreed upon, Indians living on the coast were allowed to collect salt for their own use. In celebration of signing the pact, Lord Irwin, the British viceroy, suggested drinking tea. Gandhi answered that he would drink water, lemon and a pinch of salt.[49]

Salt's "fortunes" began to change once chemistry was able to better understand its composition. For centuries people had been aware that a wide range of salts existed around them. Not all salts were the same. They, moreover, knew of a wide range of applications in which they could be used. Ancient Egyptians, for instance, could discern the difference between sodium chloride and natron. The Chinese in the middle ages were aware that saltpeter (which could be either sodium nitrate or potassium nitrate) could be used to make gunpowder. What they lacked, however, was an understanding of their individual chemical composition. Taste had been the way that earlier civilizations were able to distinguish one salt from other such as sodium chloride from magnesium chloride. In 1807, however, the Englishman Sir Humphrey Davy was able to isolate sodium, the seventh most common element, using a process of electroly-sis. Even before that, for example, in 1792, sodium carbonate, or soda, was made from mother liquor, a residual liquid that results from crystalliza-tion. Soda existed in nature, and was used extensively in the glassmaking industry. The breakthrough of being able to manufacture it, however, led to the creation of many new industries.[50]

Much of the experimentation with the properties of salt was associated with revolutionizing warfare and creating deadlier weapons. Technological breakthroughs took place with great speed. These also impacted the trans-portation networks. Canals for example were made obsolete and they gave way to the railroads. Both changes in technology and transportation affected a number of areas in which salt production had flourished, caus-ing them to lose importance.

Some of the innovations that impacted salt's importance as a strategic commodity centered around preservation. Nicolas Appert[51] in 1803, for instance, found that sealing food in a jar and then heating it would lead to its preservation. Interestingly enough, this innovation was a product of a "purse" of 12,000 francs offered to any person who could create a method to preserve food for use by an army on the march. Appert won the com-petition but could not publish his method for ten years, because it was considered information of strategic importance.[52]

In 1809, moreover, Londoner Peter Durand received a patent for pre-serving food in tin and other metals instead of glass bottles. Bryan Dorkin, an industrialist from Britain founded the first canning plant. Shortly after, the British Navy adopted this technology and used canned goods as part of its provisions. In 1830, another canning plant, in La Turballe, France, began canning sardines, resulting in the collapse of the salt-fish business in that area. Similarly, this new method of preservation was also respon-sible for the collapse of both the salt herring and anchovy industries.

Innovations in refrigeration technology eventually dealt the mortal blow to the salt fish industry. The novelty stemmed from the notion of using cold temperatures, not salt, to preserve food. Artificial refrigeration research had begun in the 1700s making slow progress. In the beginning of the 1800s, Thomas Moore, an American, built a primitive icebox in which to preserve butter. The box itself was made of wood, but inside it had a metal container to hold the butter that was surrounded by ice. He stuffed the box with rab-bit fur for insulation and his butter stayed both firm and chilled. As a result Iceboxes for domestic use became increasingly available in the 19th century, along with ice-storehouses as ice harvesting techniques evolved.

New Yorker Clarence Birdseye[53] moved to Labrador, in what is today the Canadian province of Newfoundland and Labrador, with his family to trap furs. He noticed that the fish caught by the locals instantly froze in the cold climate, a combination of ice, wind, and low temperatures. Further, when the fish were thawed, often many weeks later, they still tasted fresh. Birdseye began to experiment and found that when food is frozen very rap-idly bacteria will not develop. Birdseye envisioned the industrialization of the fast-freezing process. In 1925, he moved to Gloucester, Massachusetts, to establish a frozen seafood company.[54]

Railroads allowed for faster transportation, and more Americans were now being introduced to fresh fish, which they vastly preferred over the salted kind. By 1910, only 1 percent of fish from New England was cured with salt. Birdseye's invention proved to be a goldmine. By 1928, a million pounds of frozen food – using his fast freezing method - was sold in the United States. Birdseye sold his company before the Stock Market crashed in 1929 to a corporation that evolved into General Foods Corporation. The fast freezing method became extremely popular leading to an ever-increasing American demand for fresh tasting fish. This new market led fishing vessels to comply with changing appetites provoking a switch to fast freezing techniques onboard their vessels.[55]

There were still other significant breakthroughs that derived from the changing salt industry, one of which was development of drilling technology. This innovation alone brought forward the new age of geology, and therefore the ability to study the earth. Drilling salt domes and modern geology made it abundantly clear, ironically, that salt deposits were actually very common.

For centuries, the great powers of the day sought salt to collect revenue, to bolster trade, build empires, generate wealth, and gain geopolitical influence. Although many commodities had been valuable throughout history, salt's pre-eminence is undeniable. The desire for salt led to competition among the powers of the day for trade supremacy and wealth. Yet these efforts to control its production, impose high tariffs, and monopolize its sale created resentment, unrest, and hardship for many peoples across the globe. It was a factor leading to the revolt in the North American colonies. The dreaded gabelle poll tax imposed in France inflamed passions and inequalities[56] and was repealed only after the French Revolution, by the revolutionary legislature. Gandhi himself "rocked" the British Empire with a "pinch of salt."

Nonetheless, salt is a salient example of a strategic resource that peaked and then declined in importance due to technological innovation and an expanded understanding of geology and chemistry from the nineteenth century onward. Today, though still vital in agriculture, food, and other industries, salt is seen as just another common commodity. It is plentiful, geographically spread out, and sold on the open market. No modern nation would go to war over salt.[57]

It is a highly ironic that for purely geological reasons, natural supplies of oil and gas are often discovered at the edge of salt domes. By being impenetrable, salt helps to trap organic materials that decompose into oil and gas.[58] In this symbolic way, the one commodity passed the baton to the next, and the innovative drilling technology created for the salt industry opened new paths to the discovery of oil.

But if salt has fallen off the strategic commodity pedestal, its overthrow took centuries to accomplish. While its reign lasted it was controlled, taxed, traded, and fought over with ferocity and in ways that today seem unthinkable. Its story shares many similarities to rare earths. Rare earths, like salt, are versatile and have many important applications that are indispensable in modern civilization. Yet, unlike salt, it is not likely that rare earths can be innovated out of importance in the years to come, because they serve as enablers in a wide range of highly complex technologies that are made more efficient through their use.

## OIL

Oil offers an even more salient case study for a comparison to rare earths because it rose to pre-eminence by powering the economic miracle of industrialization that defines the modern era. In today's world, energy resources are indispensable and coveted by all nations large and small. Coal, oil, gas, and, increasingly, renewables are all vital inputs that keep global economies running. If coal was pivotal in ushering in the industrial revolution, mechanization in industry, and faster transport across land and sea, the discovery of oil was the game changer, becoming the universal fuel in the modern age.

Indeed, oil has been much more than a strategic resource. It has been the very foundation of growth and economic development from World War II onward. Until now, and contrary to the modern Cassandras who for years have predicted that the age of oil would soon be over, it has remained at the top of the strategic commodities list. Uninterrupted access to oil has been a crucial concern for all major economies because, until very recently, the major players dominating the oil market were concentrated, by and large, in a rather narrow geographic area. Moreover, interruptions had occurred because of regional unrest or all-out war, but also when oil was used as leverage in geopolitical disputes. Each time this occurred, the global economy was adversely impacted. Increasingly, the ever-growing energy needs of quickly developing countries, such as China and India, aggravated competition for this strategic resource.

While fossil fuel hegemony is expected to continue, a combination of new information and technological innovation is proving significant enough to give rise to new geopolitical tensions and realignments in a game of economic hardball over industry dominance, market share, output, and pricing, as well as a pronounced reorientation of the energy economy to include renewables. First, the breakthrough innovation of hydrofracturing has revolutionized fossil fuel extraction of oil and gas, transforming the United States into a powerful fossil-fuel producing nation that can rely extensively on its own resources to power its economy. This development has resulted in a number of significant changes. Being able to rely on domestic production the United States is now less dependent on imports and therefore, for example, seeks to become less engaged in affairs of the wider Middle East, even during a period of renewed turmoil. Furthermore, the ability to keep energy prices low has contributed to a renaissance of the US economy, and especially the manufacturing sector. The United States' newfound importance as an oil producer and non-OPEC member has stirred tensions and created an "unacknowledged" type of rivalry. Second, strides in geology have proven that oil and gas reserves are plentiful in many countries around the world, which will allow energy-consuming nations access to more suppliers, alleviating energy security fears. That is, oil and gas are now considered plentiful, at least in theory. The problem is that the extraction and production of many of these newly discovered resources require considerable capital investment and are economically viable only when the price of oil hovers around the $100 a barrel mark.

Oil prices, however, as in the case of other commodities, are cyclical, and after a long period of high prices we have now entered an era of plummeting prices: the price of oil was less than $50 a barrel in the first half of 2015; in 2016 it remained mostly in the $40 to $50 range,[59] and continued in that range in 2017. These low prices are a result of a prolonged global recession, the increase of production of oil and gas by the United States (which reached 9.5 million barrels a day in 2015) and OPEC's decision not to cut its own output to "accommodate" the additional US oil on the market. OPEC's strategic decision was to hold out for as long as possible in order to maintain what now it increasingly views as more important—that is, its market share in the global oil market.[60]

As a result of these developments, promising alternative sources of oil and gas discoveries are of no immediate economic interest. In fact, a wide range of new projects and contracts have been canceled until further notice. The countries that have historically dominated the market thus remain in central control, defending their market share tooth and nail, bracing themselves to weather the storm while prices fluctuate significantly.[61] It is important to note that those hardest hit by low prices have been OPEC members whose government budgets rely most heavily on fossil-fuel revenues. The problem is especially acute for countries like Nigeria and Venezuela that do not have sovereign funds at their disposal to help bridge the gap and have had to resort to austerity measures in their countries, leading to social unrest.

Yet this geostrategic game surrounding oil is not merely one of economics. International political realignments are taking place. China, for example, is building closer ties to the Gulf (an area where the Americans and the British have traditionally been dominant). Saudi Arabia and Russia a non-OPEC fossil-fuel producer, signed an oil-cooperation agreement in September 2016, to curb output—something the oil producers were aiming at to gain a degree of stability. Whether or not this agreement will substantially impact the price of oil in the long term remains to be seen, but it does give a sense of the power realignment between two of the most significant fossil-fuel producers.[62]

These recent developments in the oil market provide us with both a unique insight into and a useful comparison to the rise and fall of rare-earth prices that has resulted in China's continuing geographic and supply hegemony. Although rare earths in their natural forms, like oil, appear to be geologically abundant, they remain scarce as a mineable resource, and their strategic importance as a commodity has not diminished because, as we have seen, substitution has proven difficult, in effect, testing the limits of the notion that innovation is a panacea for scarcity. Moreover, just as with oil, new mining projects outside China that had seemed very promising when the prices of the elements had skyrocketed are no longer the alternatives the world thought they were because they are no longer economically viable.

There is, however, one more development that should attract our notice. The growing climate crisis now demands a change in the world's energy mix. Suddenly, other competing energy resources are being deployed to buy the world more time in which to reach carbon neutrality and offset a planetary catastrophe. Rare earths are crucial inputs in the production of renewable energy applications and will be used increasingly in the fight against climate change.

The overall importance of oil may at present seem disproportionate in comparison with rare earths; however, rare earths are crucial enablers. Minor quantities of these elements are inextricably linked to the performance of vital technological applications. Thus, there are lessons to be learned from taking a closer look at oil as a strategic commodity that should not go unnoticed when thinking about the rare-earth crisis, its aftermath, and the centrality of these elements. In what is increasingly becoming a primarily digital high-tech world, we can no longer turn back the clock to simpler technological solutions and applications.

These challenges have not been lost on China. A closer reading of the PRC's single-minded focus on securing oil and other energy resources reveals its long-term-planning approach to issues that it perceives to have strategic national importance. In the case of oil, just as with salt, China has demonstrated singular purpose. Once its domestic supplies began to decline[63] and the PRC decided to turn to a system of state-run capitalism, China's growth has been unrivaled, and its energy needs have grown exponentially as a consequence. Today, it is the world's largest consumer of oil.[64] To guarantee an accelerated speed of economic growth, China crafted a centralized and clearly mapped out energy policy[65] to ensure its energy security. It built relations with oil-producing nations, not only in the Middle East, but also in Asia, Africa,[66] and Latin America, through diplomacy and development aid,[67] using economic statecraft to bolster its advantage and to achieve its goals.

The PRC actively sought to invest in the oil industry both domestically and abroad, and to develop a more balanced energy mix to ensure that it would never be found wanting for energy resources.[68] This multifaceted approach that is the cornerstone of its energy policy[69] also helped China to identify rare earths early on as the next crucial strategic material. Control of rare earths allowed the PRC to take the necessary steps to ensure that it would become not only the leading producer (and consumer) of renewable energy, but also of the manufacturing of high-tech and military applications as well.[70] Both these areas enhance China's energy security and greatly add value to its economy.

## OIL: FROM DISCOVERY TO STRATEGIC IMPORTANCE

From the outset, oil has been a versatile resource with multiple uses. It not only powers the global economy but is the basis for a wide range of products that have become indispensable: plastics, fuels, petrochemicals. Yet oil as a nonrenewable resource would peak[71] and eventually run out,[72] and in the beginning its production was thought to be geographically highly concentrated while its uses were universal.[73]

In April 2002, Robert E. Ebel, of the Center for Strategic and International Studies, told an audience from the State Department, "Oil fuels more than automobiles and airplanes. Oil fuels military power, national treasuries and international politics." Oil is not just a commodity being bought and sold on the international market; it determines the well-being of nations, their national security, and their international power.[74]

From the early 1920s to the 1970s, though oil belonged to the countries of the Middle East, it was controlled by big European and American companies. These companies had the know-how and the capital to invest in extraction and processing, and they transported the oil to the consumer. Local governments in those countries benefited through the concessions they extended to the Oil Majors[75] drilling in their region, and yet they felt cheated of what was rightfully theirs. The ensuing bitter struggle gave back control of oil to oil-producing nations by the 1970s, and their governments gained the economic and political power to shape the modern history of the Middle East.[76] In that half-century, oil evolved into a resource of critical strategic importance as a result of rapid industrialization and reconstruction following World War II.

The world became addicted to oil, and high demand relative to restricted supply created a fragile market. Disruptions in production would cause tremors across the world economy, as the biggest oil shocks clearly showed. The industry had changed as well, as numerous new companies began taking their places on the world stage, a fact that in itself curbed the Oil Majors' power and influence. OPEC, too, played a pivotal role in this transition.

OPEC was founded, in 1960, by five countries: Iran, Iraq, Kuwait, Saudi Arabia, and Venezuela. Today, it has twelve members, including Algeria, Angola, Ecuador, Libya, Nigeria, Qatar, and the United Arab Emirates. According to its official mission statement, OPEC's objective is to coordinate and unify petroleum policies among its member countries. This is done "to secure fair and stable prices for petroleum producers; an efficient, economic and regular supply of petroleum to consuming nations; and a fair return on capital to those investing in the industry."[77]

By 2005, OPEC's net export revenues were $473 billion, double that of 2001.[78] Given the increase in oil prices, which were hovering around $100 a barrel by the end of 2011, the US Energy Information Administration had predicted that OPEC revenues would collectively break the $1 trillion mark in 2011 and in 2012.[79] OPEC may have been established in 1960, when the world oil market was dominated by the "Seven Sisters," but it became globally important after domestic production fell under the control of the member countries, and it gained a powerful say in the prices of crude oil set on the world markets. OPEC's role has been to serve as a tool of cooperation between the oil-producing member states to exert market power.[80]

OPEC proved that when a vital resource was controlled by a small number of producers who could find a modus operandi to cooperate in order to regulate their output and service the market, their power grew in manifold ways. This became painfully clear during the oil shocks of 1973–74 and 1979–80, which led to shortages, higher prices, and global recessions.

In their 2004 paper "The Effects of the Recent Oil Price Shock on the U.S. and Global Economy," Nouriel Roubini[81] and Brad Setser emphasized the repercussions of oil prices on both the US and the global economy. "Oil shocks have caused and/or contributed to each one of the US and global recessions of the last thirty years. Yet while recent recessions have all been linked to an increase in the price of oil not all oil price spikes lead to a recession." To wit:

- The 1974–1975 US and global recession was triggered by the tripling of the price of oil following the Yom Kippur war and the following oil embargo.
- The 1980–1981 US and global recession was triggered by a spike in the price of oil following the Iranian revolution in 1979.
- The 1990–1991 US recession was partly caused by the spike in the price of oil following the Iraqi invasion of Kuwait in the summer of 1990.

- The 2001 US and global recession was partly caused by
  the sharp increase in the price of oil in 2000 following the
  California energy crisis and the tensions in the Middle East (the
  beginning of the second intifada). But other factors were more
  important: the bust of the internet bubble, the collapse of real
  investment and in smaller measure the Fed tightening between
  1999–2000.[82]

The first oil shock, of 1973, began when the OPEC countries decided to enforce an embargo because of the decision on the part of the United States during the Yom Kippur War to resupply Israel's military. The embargo targeted industrialized nations such as Canada, Great Britain, the Netherlands, and the United States. It was not a uniform embargo, however. Some European countries (members of the European Economic Community—EEC) faced partial oil cuts. As a result of this embargo the price of oil quadrupled at the time, from $3 to nearly $12 a barrel. This had two major effects. Economies were hard hit; but the price hike reversed the fortunes of the oil-producing nations, which saw their revenues go up dramatically. The embargo was lifted in March 1974 after negotiations at the Washington Oil Summit.[83]

What the crisis also revealed—and which continues to provide useful lessons for major industrial actors—was the complete lack of cooperation and coordination between the major oil-consuming nations. Countries adopted approaches that were narrowly self-interested. The crisis brought back memories of wartime shortages, sending European nations into a tailspin of uncoordinated reactions to the disruption in the oil market.[84] Britain and France began by attempting to appease the Arabs in isolating the Dutch, who were being targeted in the embargo. The EEC at the time went so far as to draft a resolution favoring the Arab position in the Middle East dispute. Some of the restrictions on shipments were relaxed as a result of this resolution. Japan followed in the EEC's footsteps and weighed in on the Arab side.

While the OECD tried to help countries weather the crisis by proposing information exchanges and an oil-sharing plan, these suggestions never took off in the widespread panic. Individual countries such as France and Britain also began to pressure their own oil companies to show them preferential treatment. Furthermore, the Americans, the Germans, and the Japanese drove up the prices in the spot market by bidding up to ensure that they had access to oil.[85] It was by trying to save themselves without any coordination of efforts that the nations managed to quadruple the price of the barrel from $3 to $12.[86]

The oil companies provided a solution at the time. Instead of bending to national pressures to show their countries of origin preferential treatment, they instead aimed to "reduce shipments equally to all their customers."[87] As Robert Keohane put it in his book *After Hegemony*, "The 1973–74 crisis illustrates the severity of the dilemma of collective action when uncertainty is high and no institutions for reducing it exist. Each country followed the 'defecting' strategy of the Prisoners' Dilemma, fearing that if it failed to try to get preferential treatment for itself, it would wind up with the 'sucker's payoff' of oil shortages."[88]

The first oil crisis made it evident that it was necessary to address issues of energy security in the event of another significant disruption of the market, which could now not be ruled out. The United Stated called for an International Energy Conference in early 1974 that resulted in the creation of the IEA (International Energy Agency) by the end of the year.[89] The need for an international regime that would help nations navigate through another oil crisis was seen as particularly important because oil discords had not only economic implications but political and security ones as well.

The second oil crisis, in 1979, occurred in the wake of the Iranian Revolution that disrupted the oil sector in that country. This unanticipated event was another significant blow to industrial nations around the globe. Once again, and even though the IEA had been established to lead countries through this kind of an oil crisis, nations reacted similarly to how they had reacted to the oil crisis of 1973, scrambling to secure energy supplies and thus driving up the price.

Though OPEC increased supply to offset the aforementioned outcomes of the panic manifested by the oil-consuming countries, shortages continued and significant price hikes took place, especially after President Carter banned all Iranian oil imports. The average price rose to $15.85 a barrel. This trend was aggravated when the Iran-Iraq war broke out, further disrupting oil supplies. At the end of this period of instability, however, the oil-consuming world received a respite because new producers had come into the market, creating an oversupply and a protracted period of low prices.

The respite did not last. Prices went up again following Iraq's invasion of Kuwait in 1990. This resulted in a seven-month occupation of Kuwait by Iraq and the Allied invasion of Iraq to liberate Kuwait. Oil prices increased from $21 a barrel in July 1990 and reached $28 a barrel a month later. The price spike subsided after US military and Allied intervention calmed the markets. The next shock was even greater. Prices rose from the $30 range to $60 by August 11, 2005, and peaked at $147.30 in July 2008.

Various theories were proposed to explain the surge in prices. Some attributed it to the geopolitical instability in the Middle East. Others based it on the presumption that oil had in fact peaked and a period of decline was about to begin, predicting shortages in the future. A number of them attribute it to China's growing demands for oil. The global financial meltdown brought prices back down to a low of $32 in December of 2008. In 2009 prices began to stabilize and traded in the range of $60 to $80 until the price spiked to $105.23 dollars a barrel on April 2, 2012.[90]

The fact that the world's largest and most economically accessible oil reserves are located in the Middle East and the Gulf region poses a series of challenges for oil-importing nations, in particular the United States and its allies (the US and EU alone represent approximately half of global GDP).[91] The Middle East and Gulf region is still evolving on many levels, cultural, religious, and political, and certainly remains in flux, with an increasing number of nations now at war and falling into the category of failed states.

Access to oil has ranked as a "vital national interest" for powerful industrial nations. Behind power politics initiated by industrialized countries is a sense of vulnerability to a decision by oil producers to stop selling oil to them—temporarily or long term (embargoes, blockades etc.). Such threats have not been only theoretical but have also been carried out in the past. The 1973 oil embargo by OPEC countries was just one such manifestation. Another historical example was the US decision in 1941 to impose an oil embargo on Japan because of its occupation of China. Japan had been relying on the United States for 80 percent of its imports. The attack on Pearl Harbor has been viewed as a "countermeasure" for the US using oil as a weapon. The threat of scarcity of oil accentuates these kinds of fears.[92]

The rapidly growing economies of China, India, and other populous developing countries will continue to rely heavily on oil in the years to come—even if they invest in renewables and in nuclear to help fulfill their energy needs. This effectively means that oil-producing countries will maintain their extraordinary power. In her testimony before the Senate Foreign Relations Committee of the United States, on April 5, 2006, then Secretary of State Condoleezza Rice insisted, "We do have to do something about the energy problem. I can tell you that nothing has really taken me aback more, as Secretary of State, than the way the politics of energy is . . . 'warping' diplomacy around the world. It has given extraordinary power to some states that are using that power in not very good ways for the international system, states that would otherwise have very little power."[93]

Oil (and gas) have, time and again, been used as geopolitical tools to demonstrate power and achieve strategic pursuits. Russia, for example, has repeatedly flexed its oil-and-gas muscles as political weapons vis-à-vis European countries of the former Soviet bloc. Russia's actions have contributed to the EU's active investment in finding ways to diversify its gas supply and in building its capability in renewables. The policies of Hugo Chavez, president of Venezuela (1999–2013), also serve as an example. He used the country's oil wealth, to consolidate an anti-US front and to become an influential player in regional politics. Oil was behind Iraq's invasion of Kuwait in 1990 and the subsequent invasion of Iraq by the US and allied forces.[94] In the Gulf region, Iran after the revolution began using its oil revenue to further its position as regional player, build its nuclear sector, and spread its conservative values to neighboring weaker states.

Oil has also been at the center of other internal brutal conflicts in newer oil-producing nations. Although countries, such as the United States, heralded the emergence of new producers as a way to diversify the sources of energy imports and reduce reliance on the Gulf,[95] the conflicts that have arisen—in these mostly poor, undemocratic, oil-producing nations—have ranged from secessionist movements (Niger Delta) to all out civil wars (Algeria, Colombia, Sudan, Angola, etc.).[96]

## CHINA'S OIL POLICY

With a population of 1.4 billion and growing, and an annual economic growth rate of nearly 6.9% in 2015,[97] as announced by Wang Baoan, the head of China's National Statistics Bureau (still high but down from 9% in 2011), China needs to have easy, secure, uninterrupted, and affordable access to all energy technologies and sources.

According to the IEA, China is already the largest energy consumer, as well as the largest oil consumer, worldwide.[98] Oil imports continue to originate from the Middle East (52% in 2014), although supplies from African countries (22%) such as Angola are growing. Chinese national oil companies are actively working to diversify the PRC's supplies by investing in upstream oil projects and securing long-term contracts.[99]

Natural gas use is on the rise, but it accounted for only 5.7 percent of China's energy consumption in 2014 (see figure 3.1). The government is seeking to expand the use of natural gas in power plants.[100] As the United States and Europe redrew the map of the Middle East and fought over oil rights and contracts from the early twentieth century, China's role in the region during that period was insignificant. China relied mainly on coal and produced its own oil, especially after the discovery and development of the Daqing oil field in 1959.

China did try to stir anticolonial sentiments in the Middle East during the 1950s and 1960s while also attempting to keep the Soviet Union in check, following the Sino-Soviet split in 1960. It was not until the late 1970s, however, that China became more involved in the Gulf region, when Deng

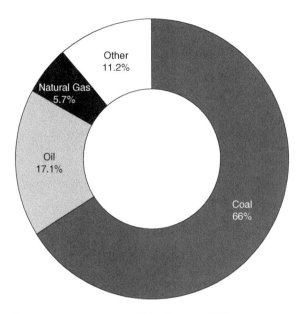

**Figure 3.1** Total energy consumption in China by type, 2014
SOURCE: National Bureau of Statistics of China

Xiaoping began to open China to the world[101] and sought to normalize relations by downplaying ideological differences. Relations with Middle Eastern[102] countries all across the spectrum were cultivated in the 1990s. Since then, China has emphasized economic ties, and its trade with the six members of the Gulf Cooperation Council (Saudi Arabia, Kuwait, the UAE, Oman, Qatar, and Bahrain) rose dramatically from $1.5 billion in 1991 to $20 billion in 2004, and then to $33.8 billion in 2005 and $196 billion in 2011.[103] China's ties with Iran are also significant. After the sanctions were lifted, President Xi was the first Chinese leader to visit the country since Jiang Zemin in 2001. His trip was meant to signify a renewed strategic partnership between the two countries which resulted in the signing of a number of accords and the agreement to expand bilateral ties and to boost trade to $600 billion in the next ten years.[104]

Post-revolutionary China had initially been of the mindset that it had to rely on its own resources in order to be safe from the outside pressures and powers seeking to curb China's growth and transformation into a world power. This thinking has today evolved and transformed. The policy that

China is developing is that of an "outward-looking oil economy."[105] After China became an oil importer in 1993, Chinese premier Li Peng[106] indicated the government's change of approach, stating that "as the economy develops and people's living standard rises, demand for oil and gas is certain to increase by large margins. While striving to develop our own crude oil and natural gas resources, we have to use some foreign resources."[107]

These energy concerns led the government to conclude the process of placing most state-owned fuel operations under the regulatory oversight of the State Energy Administration, by 1998. [108] China established two major firms in the oil industry, the China National Petroleum Corporation (CNPC) and the China Petrochemical Corporation (Sinopec). Both are significant global players in the world of oil today and are involved in both exploration and production. Furthermore, since 1982 the China National Offshore Oil Corporation (CNOOC) has been conducting offshore exploration and production.[109] These major companies are primarily state owned. As they continue to grow, they are developing their own business agendas. It remains clear, however, that the government is keen to maintain control over these companies so that their activities reinforce—among other goals—China's national strategy with regard to both resources and foreign policy.[110] This was made evident in 2011, during a reshuffle of the top executives of the three major oil companies. According to *China Security*, a journal on China's strategic development,

During the first week of April 2011, the Chinese Communist Party (CCP) reshuffled top executives of China's three major national oil companies (NOCs): China National Offshore Oil Corporation (CNOOC), China National Petroleum Corporation (CNPC) and China Petrochemical Corporation (Sinopec). On April 2, the CCP announced that Su Shulin, the former party secretary and general manager of Sinopec, would become the deputy party secretary and acting governor of Fujian Province. On April 8, the CCP revealed that Fu Chengyu, the former party secretary and general manager of CNOOC, would become chairman and party secretary of Sinopec. The CCP also announced that Wang Yilin, a deputy general manager of (and the number three official at) CNPC would become chairman and party secretary of CNOOC.[111]

This reshuffle of executives gave clear indications that the Communist Party maintained control over China's most prominent and visible oil firms. As opposed to the process of selecting CEOs typically seen in companies such as ExxonMobil and Shell, where an independent board of directors or senior management makes the appointment, the leaders of China's national oil companies are nominated essentially by the division of human resources of the CCP and ultimately require Politburo approval.

China's deep-seated worry is that when push comes to shove and its demand for energy and other natural resources outstrips supply, then the other great powers, especially the United States, will stand in the way of China's access. The Chinese perspective is that the United States, especially, has not only built a strategic dominance in the Persian Gulf but the US Navy also has control over critical sea lanes for transporting energy as well as enormous influence in the global oil industry.[112] Because China was a latecomer to the importance and long-term strategic value of foreign oil, it perceives the partnerships with key producing countries that European, Japanese and American companies have already spend decades building, as solidified. China therefore had to emphasize its presence and build its relations initially with countries where Western interests were less prominent. These areas were opportunity targets for China. It turned its attention to Russia and Kazakhstan, Yemen, Oman, and several African countries and to countries whose relations with the West were traumatized, such as Sudan and Iran.

It has, however, not shied away from seeking to also build ties to countries that both Europe and the United States already have vested interests in. China brings its extraordinary market size to the negotiating table and offers the political alternative of working with a country that has not been the focal point of tension and strife in oil-producing regions for many decades. The main mission of Chinese state-owned companies is to acquire foreign energy resources, preferably by signing long-term contracts in addition to purchasing overseas assets in the energy industry.

This approach allows China to circumvent exclusive reliance on the global oil market.[113]

Throughout the 1990s, China was making deals to secure long-term supplies and also to buy facilities in Africa and Latin America. CNOOC[114] became the largest offshore oil producer in Indonesia in 2002. China has developed a strategic oil reserve to keep itself operating for three months without any imports whatsoever. It has been emphasizing its cooperation with Russia and Kazakhstan[115] to avoid dependence on shipping routes.[116] Chinese oil companies have purchased oil fields in Canada and Peru. CNPC signed an agreement in 1995 with Japan's Marubeni Corporation for downstream joint ventures in other countries. It also entered a joint venture with an American partner and bought ninety-eight old oil wells in Texas.[117]

China does not only have vulnerabilities; it also has a number of weapons in its arsenal to draw on to help seal its energy deals. Its membership on the UN Security Council and in other UN bodies carries significant weight with other nations. Its robust defense industry makes significant arms sales to countries, such as Iraq, Iran, Sudan, Angola, and Nigeria, from which it seeks oil concessions. It has been known to take conciliatory approaches to border disputes (as it did with Kazakhstan[118]) if its energy interests so dictate. It can offer diplomatic support and assistance to the energy-producing nations with which it maintains ties in cases of disputes with their neighbors and beyond (Angola, Sudan, Iran).

From the moment that China decided that it was time to build its relations with the Persian Gulf countries, it has aimed to enhance its position in the region, exhibiting an admirable understanding of the power struggles at play. China knows that for countries in the Gulf it is becoming politically more expedient to not have to rely uniquely on the power and relationship that they have built with the United States. Without seeking to replace the United States in the region, China is asserting itself by strategically building new partnerships, interdependence with their trading partners, and molding itself quietly as the benign power that provides a healthy alternative to the entrenched status quo. China has sought to develop any and every opportunity it could find, and that is why it has unique relationships with each country in the Gulf.

In January 2012, China's premier, Wen Jiabao, visited three oil-and-gas-producing nations, Saudi Arabia, Qatar, and the United Arab Emirates. The trip was the first in two decades by a Chinese premier to Saudi Arabia and the first ever to the other two countries. China had damage control to tend to, given that in 2011, Libya's Col. Muammar el-Qaddafi—who has since been swept out of power and killed—had been offered the sale of weapons by the largest Chinese state-owned arms company to quell the revolt.[119] In fact, China maintained excellent relations with Yemen and Syria, as well as Libya and Iran.

Iran is China's third-largest supplier of crude oil, at approximately 500,000 barrels a day. So Iran plays a critical role in China's geostrategic and energy security plan. A loss of Iranian oil would not only need to be immediately replaced so as to not create a supply shock in the Chinese economy, but China has also signed multibillion dollar contracts with Iran for energy exploration and refining. Furthermore, in the period of sanctions and embargoes against Iran, China walked a fine line, sticking to its statement that the sanctions lacked the legitimacy of the United Nations while all the while keeping interests of other trading partners under consideration.[120]

US Treasury Secretary Timothy Geithner's visit to Beijing in January 2012 spoke volumes on this point, as he tried to change China's stance on Iran. In his statement, Geithner indicated, "On economic growth, on financial stability around the world, on non-proliferation, we have what we view as a very strong co-operative." Zhai Jun, China's vice foreign minister, made China's concerns especially clear, "Iran is also an extremely big oil supplier to China, and we hope that China's oil imports won't be affected, because this is needed for our development . . . We oppose applying pressure and sanctions, because these approaches won't solve the problems. They never have. We hope that these unilateral sanctions will not affect China's interests."[121]

The relationship between China and Iran, though mainly economic in nature,[122] has expanded in other areas. In July 2005, Iran together with India and Pakistan were given joint-observer status in the Shanghai Cooperation Organization, which is a regional security instrument to combat separatism and terrorism.[123]

China has moved deliberately, cautiously, but with determination to secure not only access, but also a stake in energy and natural resources all around the globe. Whether it is in the Gulf region, Africa, or Canada, Chinese firms are buying stakes and whole operations. Recently, Athabasca Oil Sands Corp. triggered an option in a 2009 deal with CNPC, which is a subsidiary of PetroChina, so the Chinese oil giant would not only be a shareholder but an owner of Canada's MacKay River oil sands project, which will be fully completed by 2023.[124] Sinopec closed a deal with Daylight Energy Ltd., which amounted to $2.2 billion.

In Canada, China has been careful in its approach. Yet with Canada's decision to become a great energy and natural resources producer, especially in its western territories, there was growing interest in investing and operating there. The Chinese picked smaller deals that did not require government approval so as not to raise heated political debates over what Canadian national interests really are. China still needs technology transfer to better access and efficiently develop its own natural resources. It wants to ensure that if its domestic production is not sufficient to meet its growing needs, it can own energy and mineral assets outside its borders. Relying on the open market is not enough. Canada's tar sands are a valuable target for the Chinese, and they are interested in building pipelines across the Pacific to ensure that they will have access to oil if they need to bypass the Strait of Malacca, which could become a center of conflict in the future.[125]

It is interesting to note that, in 2011, PetroChina surpassed ExxonMobil in crude oil production.[126] While all oil companies are looking aggressively for new finds to replace their existing wells, the Chinese have been among the boldest. The government owns 86 percent of PetroChina's stock, and China itself uses almost all the oil the company pumps. With less of an interest in the bottom line than other publicly traded oil companies and given that oil prices had remained particularly high for many years, PetroChina's mission became, above all else, to find new ways to satisfy the country's increasing demands for oil. The company has been pumping all it can from China's aging oil wells, and has acquired reserves in Iraq, Australia, Africa, Qatar, and Canada. According to the IEA, the total acquisitions by Chinese energy firms in 2009 and 2010 jumped to $48 billion (from $2 billion between 2002 and 2003).[127] More specifically according to an IEA, in the three years prior to the 2014 report, China had spent a total of "$73 billion in upstream investments and [in 2014] operated in more than 40 countries to control about 7% of worldwide crude oil output."[128] According to the *Wall Street Journal*, moreover, during the first five months of 2016, China had already overseas deals worth $110.8 billion, topping 2015.[129]

China had developed its African ties in the 1950s. At the time, a mostly isolated China turned to Africa to market its limited range of products and build diplomatic relations. Today, oil exploration is one of the key areas of Chinese investment on the African continent. China's second largest source of crude oil is Africa, which accounts for 9 percent to 10 percent of the world's oil reserves. China's largest suppliers are Angola, Sudan, the Republic of Congo, Equatorial Guinea, and Nigeria. Gabon, Algeria, Libya, Liberia, Chad, and Kenya also supply China with oil. China not only trades oil with these countries, but also iron ore and metals in exchange for its exports in machinery, electronics, transportation, and communication equipment. Sino-African trade reached $126.9 billion in 2010. According to figures cited in the *Wall Street Journal*, trade mainly dominated by commodities and raw materials reached close to $166 billion in 2014.[130]

China's policy has been twofold. On the one hand, "it has offered resource-backed development loans to oil and mineral-rich nations and has on the other hand developed special economic and trade zones with several countries in the region."[131] China named this "special relationship" strategy as its "win-win" China-Africa cooperation policy.[132]

There is no denying the fact that China is interested in strengthening its ties with African nations. Its policy of nonintervention into a country's internal affairs frustrates the world community in cases of egregious violation of human rights, causing tensions to rise. There has been growing resentment as well when China's nontransparent economic exchanges thwart efforts to bring accountability and democratization.

While criticism of China's noninterference policies and investment even in conflict regions grows within the international community,

China's heightened positive status in Africa may also begin to outgrow its grace period, as its presence on the continent solidifies. African civic groups have begun to wonder, if China buys all their resources, what will this in fact do for Africa's economy and its people. Will they become just a market for China to sell its manufactured—added value— products? "The key must be mutual benefit," Trevor Manuel, the finance minister of South Africa told a group of Chinese officials, "otherwise we might end up with a few holes in the ground where the resources have been extracted, and all the added value will be in China."[133]

China's policies will be tested over time, and Beijing may find itself becoming increasingly criticized by unions, NGOs, and civil society. So far, however, with all the shortcomings and the potential backlash, China continues to strengthen its ties across the continent[134]—a fact that causes its rivals to worry. At present, China has found a way to make its presence and interests less heavy-handed and controversial in comparison to traditional Western powers, which share a long and fraught colonial past or an interventionist record. Western countries have been exploiting Africa for centuries. As China's policies unfold, it gives the West an important opportunity to evaluate Chinese geostrategic and economic intentions as the race to control limited resources heats up.

While there are those who believe that the West has perhaps overestimated the orchestration between Chinese companies investing in these regions and the government that oversees them, the PRC continues to be governed by a highly centralized system, and though it may begin to slowly lose its overriding control on company priorities abroad, it still has the power to make or break operations. Its policies are therefore far more crystallized and targeted than that of other nation-states and leading powers. Its geostrategic moves are carefully planned to ensure China's energy and resource security so that its domestic development is unhampered. Its one-party system and state-run capitalist economic model enable it to participate in the global economic arena, while being able to make critical centralized decisions that its rivals—many of them mature democracies—cannot without the consensus of numerous actors, ranging from the public, NGOs, industry, media, financial and political institutions.[135]

China's approach to securing oil speaks volumes as to how it thinks of strategic resources. Although oil has been the strategic resource powering the industrial miracle, rare earths are essential in the new high-tech digital world that not only needs them in its transition to renewables, but also in a wide range of applications that drive our new routines and information revolution. It would perhaps behoove us not to ignore telling acts of realpolitik merely by looking exclusively at the market size of each commodity. We may still need millions of barrels of oil a day to keep our world moving, but those small quantities of rare earths are just as vital to make our most prized inventions and applications efficient, compact, and high performing.

# How China Came to Dominate the Rare Earth Industry

Improve the development and application of rare earth, and change the resource advantage into economic superiority.

—CHINA'S PRESIDENT JIANG ZEMIN, *1999*[1]

The United States was initially the leader in producing and trading rare earths, and in finding ever advanced technological uses for them. The discovery of rare earths at Mountain Pass, California, in 1949, had been an important event for the US science community. Russia and the United States, the two world superpowers, were in the process of creating a balance of fear through the threat of nuclear weapons. To achieve this, however, both countries needed uranium. It was a radioactive signature associated with a mountain outcrop that led to the discovery of Mountain Pass. Prospectors thought they had "struck" uranium, and after analyzing the materials, laid claim to the deposit. The ore they had discovered was identified as flouro-carbonate bastnaesite, and the radioactive material was thorium (in small amounts) with only very minor traces of uranium.[2] By 1953, the mine had come to be owned by the Molybdenum Corporation of America, which had begun producing bastnaesite. It was initially designed for the separation of europium, which quickly became an important element in making color televisions.[3] Molycorp, as it was known, also extracted lanthanum, cerium, neodymium, and praseodymium, and scientists quickly began to discover new uses for these additional materials. The Molycorp mine dominated rare-earth production and exports for the next few decades, until China began to discover the full potential of its own resources.

New applications that required rare earths led to a growth of demand. One such application was the use of rare earths in mischmetal (alloy of rare earths) used extensively for the Alaskan oil pipeline. In the late seventies, prices for the elements increased significantly in line with inflationary pressures in the United States. Double-digit inflation after 1978 in combination with high energy prices pushed rare-earth prices upward, in line with operational-cost increases that impacted the mining industry.

Prices stabilized as a result of the US economic recovery. One curious exception at the time was the price of scandium, which was mainly produced in the Soviet Union. In 1984, the USSR ceased exports of scandium on account of "laser research." Its price, according to a US Geological Survey report by James B. Hedrick, skyrocketed to $75,000 per kilogram.[4] This anomaly ended when US production of scandium was brought online.

Other exogenous factors also impacted the rare-earth market. New environmental legislation that reduced the lead content in gasoline dampened the demand for the elements in petroleum fluid-cracking catalysts, where rare earths were used extensively. The result was a sharp decline in prices. Production in the United States was cut to offset the price decline, which resulted in supply shortages and caused prices to rebound. Overall, rare-earth prices were volatile in the 1980s and 1990s. They became dependent on the type of rare-earth element that was in demand. High-purity products, such as neodymium and dysprosium, began to see price increases because there was a growing demand for neodymium iron boron magnets in which the two elements were used.[5] The next impact on rare-earth prices was caused by China's dynamic entry into the market.

## CHINA REALIZES THE IMPORTANCE OF ITS RARE EARTH RESOURCES

Ding Daoheng, a well-known Chinese geologist, discovered a wealth of rare-earth deposits in Bayan Obo, in inner Mongolia, in 1927.[6] A few years later, it was confirmed that the deposits contained bastnaesite and monazite. The Chinese built a mine in the 1950s and began recovering rare earths in the process of producing iron and steel.[7] In the 1960s, China also discovered bastnaesite deposits in Weishan County, Shandon, and in the 1980s, more basnaesite in Mianning County, Sichuan. The recovery of rare earths, especially from Bayan Obo, became a major priority for the Chinese, and they hired technical personnel to help develop and advance their methods of recovery. They invested heavily in the research and development of rare-earth technologies. Production levels increased with growing demand. Between 1978 and 1989, China averaged an increase of 40 percent annual production, thus becoming one of the world's largest producers.[8] During the 1990s, its exports grew rapidly, causing prices to plummet, a strategy that either put competing companies out of business or drove them to greatly curtail operations.

Bayan Obo is the world's largest REE resource.[9] It is estimated that the total reserve of iron in that region stands at 1.5 billion metric tons, with an average grade of 35 percent. The same deposit is estimated to include 48 million tons of rare-earth oxides, with an average grade of 6 percent. It contains close to one million tons of niobium, with an average grade of .13 percent. Considered the most valuable rare-earth production site in the world, in 2005, it accounted for 47 percent of the total rare-earth production of China, and 45 percent of that of the world. In addition, the rare earths in Bayan Obo occur primarily in monazite and bastnaesite, and contain very high REE content (6%) and extremely high LREE to HREE ratios. [10]

In 1990, the Chinese government declared rare earths a "protected and strategic mineral."[11] This was clearly a strategic move on the part of a state that had begun to understand the potential that the rare-earth industry had for China. Since then, China has sought effective ways to increase centralized control over the industry, create a higher market value for the elements, build supply chains inside China, develop technical knowhow, and attract high-tech companies using rare earths to manufacture final products inside the PRC.

From the moment China declared rare earths to be a "protected strategic material," it meant that foreign investors could participate in rare-earth smelting and separation only as part of a joint venture with Chinese firms. Foreign investors were also prohibited from mining rare earths. Smelting and separation projects similarly required Chinese state approval. Joint ventures, moreover, needed the approval of the Chinese State Development and Planning Commission as well as that of the Ministry of Commerce.[12]


## CHINA FIRST CAPTURES THE MAGNET INDUSTRY

With regard to the supply chain, China sought first to capture the magnet market—as samarium became a key ingredient for supermagnets made of samarium cobalt. Today, magnetic technology is perhaps one of the most important uses of rare earths both commercially and militarily. Permanent magnets that utilize rare earths not only provide greater magnetic power, but they also can be much smaller in size. The issue of size is critical in applications like computers. The samarium cobalt (SmCo) magnet and the neodymium-iron-boron (NdFeB) magnet are the two leading REE magnets on the market. They are particularly useful for military applications such as missile-guided systems because of their thermo-stability.[13]

The neodymium-iron-boron (NdFeB) magnet was introduced in the 1980s. The story behind the NdFeB magnet is revealing of the Chinese modus operandi with respect to controlling the REE industry and an important indication that China attempted to corner the market for rare earths by design. When these magnets were created, two companies, General Motors and Hitachi, acquired patents. GM patented the "rapidly solidified" magnets, and Hitachi the "sintered" magnets. GM then proceeded to establish a company to produce the magnets for its vehicles. It was named Magnequench. In 1995, two Chinese groups[14] joined forces with a US investment firm and attempted to acquire Magnequench. The US government approved the acquisition after a review, and the deal was allowed on condition that the Chinese agree to keep the company in the United States for at least five years. The day after the deal expired, the company shut down its US operations; employees were laid off, and the entire business was relocated to China.

The deal was a strategic mistake on the part of the United States, because when the business left, so did the technology. In 1998, 90 percent of the world's magnet production was in the United States, Europe, and Japan. Within a decade, the bulk of the magnet industry had moved to China. Today, China continues to try to corner the magnet industry. Chinese producers have turned their attention to Japanese companies, which hold the majority of the rare-earth magnet patents when China is in fact the producer of nearly 90 percent of the global supply. In 2014, for example, seven Chinese rare-earth companies took Hitachi Metals to court in the United States, claiming that after its patent expired Hitachi was creating unfair market barriers preventing them from exporting independently and had violated international patent law.[15] This is another indication that China seeks to add value to its economy in line with the "Made in China 2025" targets that the government has set in motion to comprehensively upgrade Chinese industry. The plan as it has been described by the State Council aims to raise domestic content of core components and materials to 40 percent by 2020 and 70 percent by 2025. It calls for an emphasis on green development and the use of innovation, emphasizing quality over quantity.[16]

## SEEKING TO BUY INTO OTHER RARE EARTHS OUTSIDE CHINA

By the same token, China has consistently attempted to monopolize REE resources worldwide. It ventured to acquire Molycorp and the Mountain Pass Mine. From 1978 onward, the company was owned by Union Oil Company of California (UNOCAL), a major American petroleum explorer and marketer. In 2005, the China National Offshore Oil Corporation (CNOOC) submitted a bid of $18.5 billion cash to purchase UNOCAL. The Chinese company outbid Chevron by a half-billion dollars.

The CNOOC bid raised concern in the United States about energy security and the deal did not go through.[17] During the heated political debate over the issue, arguments for the need to defend national security prevailed. James Woolsey, former director of the Central Intelligence Agency under President Clinton, weighed in at a hearing of the House Armed Services Committee stating emphatically, "This is a national security issue. China is pursuing a national strategy of domination of the energy markets and strategic dominance of the Western Pacific."[18] Little attention was paid to the fact that had the deal gone through, the Chinese would have acquired Mountain Pass as well, solidifying their monopoly over REEs worldwide.

China also attempted, in 2009, to acquire a 51 percent stake in the Lynas Corporation, which is in possession of the Mount Weld mine in Western Australia, considered the richest deposit of rare earths outside China. The Chinese company attempting the purchase, China Nonferrous Metal Mining Company (CNMC), terminated its $505 million bid for a controlling stake in Lynas, citing the stringent demands of Australia's Foreign Investment Review Board, which had stipulated that the CNMC reduce its ownership share to below 50 percent and hold a minority of seats on Lynas's board.[19] The Chinese did not welcome the decision. "For a long time, China has had an open policy when it comes to foreign companies investing here. We hope other governments can take the same position when it comes to Chinese firms," Foreign Ministry spokeswoman Jiang Yu said in September 2009.[20] Once again, had this deal materialized, the world's dependence on China would have been nearly complete.

The minutes from the review board meeting, on September 23, 2009, gave voice to these strategic concerns. The sale of a controlling stake in Lynas was considered to be against Australian national interest.[21] "We have concluded that they would not be able to exclude the possibility that Lynas's production could be controlled to the detriment of non-Chinese end users," the minutes show. That would have been "inconsistent with the government's policy of maintaining Australia's position as a reliable supplier to all our trading partners and hence potentially contrary to national interest."[22] The Jiangsu Eastern China Non-Ferrous Metals Investment Holding Co., however, did acquire a 25 percent stake in Arafura Resources, which owns the Nolans Bore mine in Northern Australia.[23]

Nonetheless, China continued to consolidate and strengthen its dominance over the rare-earth industry, a strategy that progressively led to the production of permanent magnets, oxides, and alloys moving there as well. This relocation of production to China resulted in the United States giving up its position as the leading researcher in the field of REEs.[24] The erosion of technical expertise is viewed as even more serious than the question of resumption of production in the United States, because China dominates all the rest of the steps in the rare-earth supply chain.

Though China made efforts to strategically control its power over the REEs and their applications, it thought that prices were too low and failed to reflect the scarcity of the resources and the damage their extraction and processing caused to the environment. Furthermore, in the first decade of the twenty-first century, the PRC emphasized the goal of developing downstream industries within the country and also promoted the goal of high-tech manufacturing by the Chinese. These targets were reflected in the twelfth five-year plan. Announced in May 2011, the list included such downstream industries as magnets, phosphors, hydrogen storage materials, and abrasive polishing materials. This twelfth five-year plan contained ambitious targets for not only improving energy efficiency and reducing carbon emissions, but also for investing and transforming China into the leading producer of renewable energy. Given the Chinese decision to diversify the country's energy mix by including vast amounts of renewables, rare earths have now become a key ingredient for the success of their green-energy applications. "Rare earths are the vitamins of modern industry and they are China's 21st century treasure trove of new materials,"[25] Wang Min, vice minister at the Ministry of Land and Resources said in 2011.

Even now, after the rare-earth crisis, the most recent thirteenth five-year plan continued to build on these priorities—that is, conservation, environmental governance, protection and restoration of ecosystems, emissions control, the accelerated shift to renewable energy, and a further emphasis on an innovation economy.[26] According to the 2017 REN21 (Renewable Energy Policy Network for the 21st Century) report, China was already the global leader for new wind power installations in 2016.

Asia overall represented about half of the added wind capacity. Moreover, while wind power installation expanded to new markets globally, Europe and North America accounted for most of the rest of installed capacity in 2016.[27] Already, China is the global leader in the solar sector.[28] In fact, it also now boasts the top capacity of power generation from wind as well.[29] An earlier report by the Chinese Renewable Energy Industries Association (CREIA), the Chinese Wind Energy Association (CWEA), and the Global Wind Energy Council (GWEC), clearly indicated the rapid growth of wind power installation over the last several years in China.

> The Chinese wind industry installed 16,089 MW in 2013, an increase of 3,130 MW over 2012, for annual market growth of 24%. At the end of 2013, the cumulative installed capacity in China was 91,413 MW, an annual market growth rate of 21%. In 2013, wind power generated 134.9TWh of electricity, making wind the third largest power generation source in China after thermal power and hydropower, providing 2.5% of China's electricity. This is less than the EU's 8%, but an increase of 25% from 2.0% in 2012.[30]

In 2016, the total installed wind-power capacity in gigawatts for China was 168.7 (GW), compared to a total of 153.7 GW for the EU and 82.1 GW for the United States.[31] China's solar capacity at the end of 2016 stood at 77.4 gigawatts (GW). The new domestic priorities of the PRC's five-year plans for the future speak to concerted government intervention in the rare-earth industry to build the particular high-growth economic sector of renewables and high-tech applications.[32]

## CHINA SEEKS TO ADD VALUE TO ITS RARE-EARTH INDUSTRY: QUOTAS, TAXES, AND THE SUPPLY CHAIN

Already, in August 2009, there was a draft report from China's Ministry of Industry and Information Technology indicating that exports would be banned within the next five years. This alarmed those in both military and commercial industries that were dependent on the Chinese. Given the potential resource scarcity that suddenly reared its head in the international rare-earth industry, there was renewed interest in exploring new REE supplies in other areas of the world. A handful of Canadian mining companies began doing just that in South Africa, Brazil, and the United States while also moving forward with existing projects.

"There has been increased interest to look into ways to mine rare earth out of China, especially given the protectionism China is applying to its resources," said Frederic Bastien, an analyst at Raymond James in answer to a question asked by the *New York Times* in September 2009.[33] Canadian companies looking for rare-earth resources outside China included Great Western Minerals Group, Rare Element Resources, Avalon Rare Metals, and Neo Material Technologies. Though the signs of China's intentions were clear in 2008, it was in July 2010, after China drastically reduced exports and was accused of withholding shipments to Japan over the two countries' geopolitical dispute in September, that the surprise over the extent of the disruption fully registered. This reaction then shook the rare earths industry and triggered an unanticipated and exorbitant rise in REE prices through 2011.

On July 9, 2010, Bloomberg News, for example, raised an alarm, "China, the world's largest rare-earths producer, cut export quotas for the minerals needed to make hybrid cars and televisions by 72 percent for the second half of the year, raising the possibility of a trade dispute with the U.S."[34] Looking more specifically at the figures, the overall reduction of quotas of rare earths from 2005 to 2010 were over 50 percent (see Table 4.1).

The abrupt reduction in the allowed quotas of REEs from China also accelerated changes by the entire chain of the REE industry and gave rise to concerns that had political implications encompassing issues of national security.

In 2011, exploration projects continued to multiply, as did investment and interest in rare-earth projects. According to the US Geological Survey,[35] in 2012, economic assessments were ongoing in North America at Bear Lodge in Wyoming; Diamond Creek in Idaho; Elk Creek in Nebraska; Hoidas Lake in Saskatchewan, Canada; Kipawa in Quebec, Canada; Lemhi Pass

*Table 4.1.* Export Quotas 2005–2010[a]

| Year | Tons of Rare Earth Oxides | Annual change |
|------|---------------------------|---------------|
| 2005 | 65,580 | – |
| 2006 | 61,070 | −6.9% |
| 2007 | 59,643 | −2.3% |
| 2008 | 49,990 | −16.2% |
| 2009 | 48,155 | −3.7% |
| 2010 | 30,259 | −37.2% |

[a]SOURCE: USGS

in Idaho-Montana; and Nechalacho (Thor Lake) in Northwest Territories, Canada. In other locations globally, economic assessments took place at Dubbo Zirconia in New South Wales, Australia; Kangankunde in Malawi; Mount Weld in Western Australia, Australia; Nolans Project in Northern Territory, Australia; and Steenkampskraal in Western Cape, South Africa.[36]

The three quota systems that China implemented vis-à-vis the rare-earth industry included the production quota issued from 2006 by the Ministry of Land and Resources, the smelting and separating quota issued by the Ministry of Industry and Information Technology that went into effect in 2010, and the export quota issued by the Ministry of Commerce. The latter was removed in 2015 in compliance with the WTO ruling (see below in section China and the WTO).

## QUOTAS, ALLOCATIONS, AND THEIR REPERCUSSIONS ACROSS THE BOARD

On December 27, 2011, the Chinese Ministry of Commerce announced its first set of rare-earth export quotas for 2012. This first round of quotas was set at 24,904 tons.[37] The Ministry issued separate quota allocations for light rare earths and medium/heavy rare-earth products. For the first time, rare-earth companies were separated into two groups, one made up of those with confirmed allocations and the second group with provisional allocations. The criterion for being placed in the first group was being able to demonstrate progress toward implementing new pollution control regulations. Those with provisional quotas would receive them only if they met various requirements by July 2012. Failure to do so would mean that their quotas would be reallocated to compliant companies. The quota of 24,904 tons of rare earths announced by the government was revised in May 17, 2012 to 25,150 tons and represented 80 percent of the allocations for 2012. This was an indication that the total for 2012 would be only slightly higher than the 2011 quotas, which had stood at approximately 30,996 tons. It is worth underscoring that 87.5 percent of the 2012 quota was made up by light rare earths, which are the most abundant elements and considerably lower in price.

"On December 27, 2012, the Foreign Trade Division of the Chinese Ministry of Commerce (MOFCOM) announced the first round of allocations of rare-earth export quotas for 2013. A total of 15,499 t of export quotas was allocated in this first round, comprising 13,561 t of light rare-earth (LRE) products and 1,938 t of medium/heavy rare-earth (M/HRE) products."[38] This was the second time that the allocations had been broken into the categories of light and medium/heavy rare earths. The first round of announced quotas for 2013 was considerably smaller than that of 2012 but the final allocation ended up being approximately 30,999 tons. One interpretation of this was that China would adjust its final quota in a way that would maintain stable price levels and control over the global supply. Moreover, in 2014, the proportion of M/HRE to total allocations was 11.8 percent. This compares to 11.7 percent for 2013 and 12.5 percent for 2012, the first year that the quotas were split in this way.

Furthermore, discussions were underway for the initiation of a central-government policy to attract processing plants to China, creating a more profitable downstream processing sector by reaping added value and by gaining technological expertise in the process. Consolidation,[39] moreover, became a critical goal in order for the Chinese government to restrict rare-earth mining operations to state-owned enterprises. The Baotou Mine has been the center of the consolidation scheme, and the plan was for Baotou Steel[40] to have exclusive rights to mine in the region. At the time, *China Daily* wrote that the central government has been planning to reduce rare-earth mines from 123 to fewer than 10; and processing firms, from 73 to 20.[41] In the most recent announcements, of 2015, it was reported by the Xinhua News Agency that China's natural resource ministry had declared that "plans consolidating the rare industry into six firms have been approved . . . The plans involve miners and processors in the industry consolidating under six firms—China Aluminum Corporation, Xiamen Tungsten Co Ltd, Inner Mongolia BaoTou Steel Union Co Ltd, China Minmetals Corporation, Ganzhou Rare Earth Group Co Ltd and Guangdong Rare Earth Industrial Group Co Ltd."[42]

The first batch of mining quota announced for 2016 was a combined 52,500 metric tons of rare earths of which medium/heavy rare-earth minerals amounted to 8,950 metric tons.[43] The six Chinese companies that are owned by the state produced 99.9 percent of China's rare-earth production quota for the first half of 2016 according to the Ministry of Industry and Information Technology.[44]

Five of these Chinese companies are listed and have produced 74 percent of the production quota for the first half of 2016. While their market value is estimated at approximately US $23.4 billion, this number does not accurately reflect the market value for the rare-earth industry because these companies do not produce rare earths exclusively but other materials and products as well (such as steel and aluminum). Moreover, the value of the other ten global rare-earth companies combined stood at $409 million. Given that both Molycorp and the Great Western Minerals Group have now gone bankrupt, this estimate should be readjusted downward. Lynas, which continues to operate, had a market cap of $145.6 million in October 2016.[45] Arafura, the second mine in Australia in which the Chinese have a large stake had a market cap of approximately $25.1 million, also in October 2016.[46]

These figures may not seem significant enough to warrant such focus on rare earths. Nonetheless, the companies for which rare earths are indispensable, such as Apple and Samsung, boast a market capitalization of $614.6 billion and $216.2 billion respectively. Even Xiaomi, the Chinese smartphone producer, is now estimated to be worth $50 billion.[47]

China's plans for the rare-earth industry also included the implementation of a unified pricing mechanism. This action was aimed at cracking down on illegal mining and at stabilizing the market. China began requiring separator companies to produce documentation that they were buying legal feed. Illegal feeds[48] may be bought at much lower prices, but companies with high standing in China would not risk their positions and would only dare to buy rare earths legally.[49] Finally, China is said to have created a large stockpile of strategic reserves in the north of the country.[50] More specifically, on August 5, 2014, Bloomberg reported that China had in fact "bought 10,000 metric tons of rare earths" for its stockpile.[51] According to Peng Bo, an analyst at China Merchants Securities, whom Bloomberg quotes, "China is facing imminent pressure to abolish the export quota, so stockpiling is part of the policy reaction to help prop up prices and keep more of the resources at home for future use."[52] According to Huan Chen, an analyst at Beijing Antaike Information Development Co., the government bought the rare earths at higher-than-market prices and are holding onto them in anticipation of increases in internal demand from domestic industry in the future.[53]

In 2012, Mark A. Smith, then the CEO of Molycorp, made some interesting observations on his blog about the 2012 REE quota announcement by the Chinese. In his view,

> China's consolidation of its rare earth separations companies enables it to exercise much tighter control on what ultimately gets produced, consumed internally and exported . . . and allows for more effective control of what it considers "illegal" production. All this points to future constraints on global rare earth supply out of China.[54]

He also underscored the fact that, by withholding quotas to twenty companies until they complied with the new environmental regulations, China demonstrated its seriousness about cleaning up the environment. He did, however, add that the new requirements would increase production costs in China while also putting "pressure on China's ability to increase its own production in line with increasing rare earth demand in China."[55]

Additionally, Mark Smith pointed out that the novelty of the announcement lay in the way the Chinese government categorized REEs into "light" versus "medium/heavy" for purposes of the split in export quotas. This was particularly interesting since, if an REE changes category from light to medium/heavy, it means that it will be in smaller supply, given that quotas in the heavy category were always lower.

Smith used didymium, which is a combination of neodymium and praseodymium, and itself a critical element in the production of high-powered, permanent rare-earth magnets, to illustrate the point. If didymium were to be placed in the new middle/heavy category with tighter quotas, it could result in supply constraints. According to Smith, "Processors in northern China will use their heavies export quotas for didymium exports, while processors in the South will not have enough heavy quotas and will need to seek to purchase those from the North."[56] Finally, he indicated that supplies of terbium and dysprosium would be tight. Because of these developments, Molycorp, according to Mark Smith, would be focusing on substitution in downstream technologies that rely less on these scarcer rare earths. For companies such as Molycorp those higher prices initially provided space in which to grow and to look for ways to remain competitive in the rare-earth industry while China turned its attention inward. What initially looked like a hopeful scenario has today been proven an illusion inasmuch as Molycorp has gone bankrupt.

China's goals in mapping out this strategic approach to its rare-earth industry can be viewed as twofold, with the underlying aim of adding value to this important resource. First, it can be argued that China sought to ensure that it could service its domestic REE needs and Chinese consumers at prices lower than those exported. Second, that it also aimed to continue to provide access to international companies that would move and maintain their manufacturing facilities in China. These companies would be required to pay more than Chinese consumers, but prices would still be lower for them than for the rest of the world. With such focused domestic priorities, international consumers would need to find other sources for purchasing rare earths.[57]

In 2011, prices for rare earths went through a correction. From 2012 and the beginning of 2013, there has been a period of market fluctuation following the initial period of alarm. Reasons for this include the economic downturn, the promise of new projects, and the continued high smuggling rates of rare earths from China. Nonetheless, prices remain substantially higher than they were before the crisis and China began adjusting its output to stabilize them.

The issue of price volatility was immediately addressed by industry leaders. In 2011, the Lynas CEO at the time, Nicholas Curtis, spoke at the company's general meeting in Australia saying that the 2010–2011 price explosion for rare earths had given rise to extraordinary media attention. The high prices, he conceded, helped to create investor interest in the sector. Nonetheless, he had to admit that these prices could not have been sustained by the industry. He went on to explain the correction, its aftereffects, and the industry's prospects.

> Since August 2011 there has been a substantial retracement in Rare Earths prices and a consequent readjustment of equity market valuations in line with a much more sober global economic outlook. We believe this price retracement is healthy for the industry. Prices are still very satisfactory but much more sustainable for our customers. Our job is to grow the overall market for Rare Earths. This is difficult at unreasonably high prices. In fact, earlier in the year a number of upstream industrial recycling processes were implemented and this has led to a one-time reduction in demand for Rare Earths. However, with these process changes now in place, the industry is now poised for renewed growth in demand and we are committed to growing with our customers by providing Rare Earths at prices that are sustainable for both customers and suppliers alike.[58]

Figures 4.1, 4.2, and 4.3 show the price fluctuation of a number of rare earths and rare earth oxides.

The steep price hikes gave both industry and the government[59] much cause for concern, especially given the numerous reports that

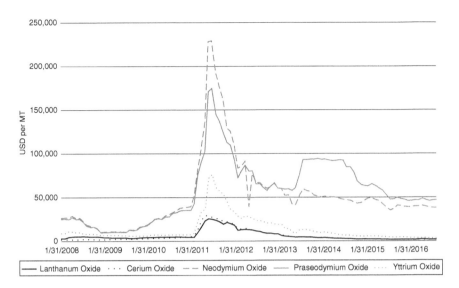

**Figure 4.1** Price changes of select rare earth oxides 2008–2016
SOURCE: Rare Earths Prices, January 31, 2008–January 31, 2016, via Bloomberg LP, accessed September 28, 2016

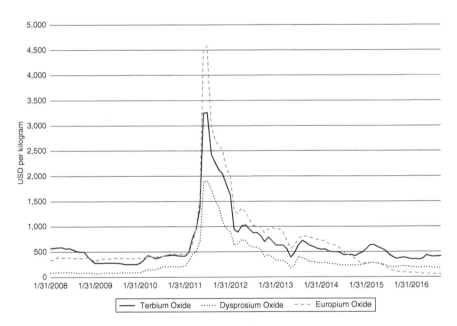

**Figure 4.2** Price changes of select rare earth oxides 2008–2016
SOURCE: Rare Earths Prices, January 31, 2008–January 31, 2016, via Bloomberg LP, accessed September 28, 2016

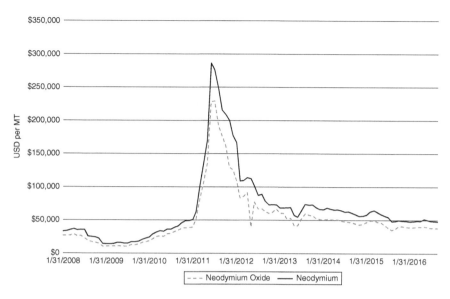

**Figure 4.3** Price changes for neodymium oxide and neodymium 2008–2016
SOURCE: Rare Earths Prices, January 31, 2008–January 31, 2016, via Bloomberg LP,
accessed September 28, 2016

raised alarms for at least five of the elements that were vital for clean-energy applications—dysprosium, neodymium, terbium, europium, and yttrium.[60] These five elements are used in magnets for wind turbines and electric vehicles, as well as phosphors, in energy-efficient lighting.

China attempted to address international concerns. While conceding that prices had significantly increased, a white paper published by the government in 2012 argued that there had been, for many years, a "severe divergence between price and value."[61] It stood by its assertion that it was time for China to protect its reserves and enforce strict environmental regulations for extraction and processing.

## THE WORLD REACTS TO THE REALIZATION OF THE GROWING SCARCITY OF REES

It took an aggressive move on the part of the Chinese for the international community and world industry to fully grasp what dependence on one supplier of rare earths might mean in the high-tech race. As Cindy Hurst put it, "The world was seemingly asleep as China grew to become a goliath in the rare-earth industry. It took the rest of the world nearly 20 years to suddenly wake up to the realization that the future of high technology could be in the hands of this one supplier."[62] One might find particularly surprising the extent to which the world seemed to have been unprepared for these developments and unable to see their stark dependency. In hindsight, closer scrutiny of Chinese policies should have made China's intentions and goals more readily apparent.

In the initial critical period in 2010, the United States, the EU, and Japan—all primarily dependent on accessible and affordable rare-earth supplies—joined efforts to address the situation. Each of these industrial powers designed and implemented a number of internal domestic strategies and policies to respond and adjust to the challenges caused by the supply disruption and the realization of China's inordinate market power. Two significant initiatives, however, stood out from the rest. First, the US, the EU and Japan filed a complaint against China with the WTO; second, they began joint collaborations through a series of trilateral workshops[63] to work on substitution, diversification, conservation, reuse, and recycling as a way to lessen their dependence on Chinese rare earths.

Both initiatives, however, were undertaken with considerable delay. The first of three trilateral workshops took place one year after the crisis had erupted, starting in October 2011. The case against China at the WTO was filed in 2012, almost two years after the height of the crisis. The case brought against China with the WTO raises questions about the timing, intent, and efficacy of the complaint itself, when it is a well-known fact that WTO disputes require from one to three years to be settled. Why did the United States, the EU, and Japan resort to tools of economic statecraft so long after the fact? In 2012, when the affected parties decided on this plan of action, prices had already dropped significantly from their all-time high in 2011. Dysprosium oxide, for example, which had reached $1,903/kilogram[64] in July 2011 was being sold for approximately $627/kilogram in February 2012, and even for less than $400/kilogram by December 2012.[65] What kind of message did the big industrial powers intend to send China by proceeding in unison and with such fanfare if the problem was short-lived? As expected, the WTO case produced a final verdict in 2014, four years after the beginning of the crisis. It was only on May 1, 2015 that China eliminated the controversial export duties to comply with the verdict.[66] Ahead of that date, in April 2015, the Ministry of Finance announced a resource tax on rare earths based on "sales value instead of production quantity."[67] Accordingly, taxes on light rare earths were set at "11.5%, 9.5%, and 7.5%, respectively in Inner Mongolia Autonomous Region, Sichuan Province, and Shandong Province, while for medium and heavy rare earths it is generally set at 27%."[68]

Before taking a closer look at the most prominent, but delayed initiatives taken against China by the United States, the EU, and Japan, it might be helpful to give an overview of how each of these major industrial actors responded internally vis-à-vis the rare earths crunch. They clearly chose from a predictable array of options that were at their disposal attempting to weather the storm, vocalize their dissent, and develop a level of resilience. Looking back more closely, however, it becomes apparent that many of the initiatives either fizzled out or never came to pass. Part of this can be explained by the different regulatory traditions that impacted the focus of their strategies vis-à-vis the formulation of a critical minerals policy. Nonetheless, this shortsightedness may seem all the more perplexing since all three powers remain particularly dependent on rare earths for their high-tech, defense, and green applications.

Europe's twenty-first-century ambition, for instance, has been to complete its transformation into a knowledge society, an innovation society, and an inclusive low-carbon economy. To achieve these objectives, Europe has identified specific targets for diversifying its energy mix with renewables and has increasingly invested in new green technologies. These goals have been in line with a political decision to lead the fight against climate change through international cooperation and to exchange best technological solutions and best practices, as well as to champion the development of legally binding agreements.

These objectives have allowed Europe to deliver a new paradigm for growth in a zone that has seen its economy contract since the 2008 global economic downturn and the prolonged debt crisis that has ensued. The "green economy" has provided Europe with an area for new growth that is also aligned with the desires of its citizens for climate action. Manuel Barroso, the president of the EU Commission in 2010 had argued, "The crisis wiped out years of economic and social progress and exposed structural weaknesses in Europe's economy. In the meantime, the world is moving fast and long-term challenges—globalization, pressure on resources, ageing—intensify. The EU must now take charge of its future. We need a strategy to help us come out stronger from the crisis and turn the EU into a smart, sustainable and inclusive economy delivering high levels of employment, productivity and social cohesion."[69]

To achieve this vision and develop its high-tech sector, however, Europe has had to increasingly rely on imports of many critical raw materials. Sustained availability of these minerals and metals is critical for its economy, and when the rare-earth crisis erupted, this issue became a pronounced concern. In the words of Antonio Tajani, commissioner for industry and entrepreneurship in 2011, "Without assured access to critical materials, the deployment of European cutting-edge technologies will not be possible. European companies need to have a secure, affordable and undistorted access to raw materials. This is essential for industrial competitiveness, innovation and jobs in Europe."[70]

European output of metallic minerals has been low, making Europe particularly vulnerable to disruptions and increasingly reliant on a global matrix of supply chains. According to the EU Critical Raw Material Report of 2014, Germany's contribution to the global critical raw materials supply was 1 percent, while for France and Italy the contribution was 0 percent.[71] Already in 1975, the European Economic Community (EEC), as it was then called, had begun to address the need for uninterrupted access to important materials and it had drafted a raw materials report. This report pointed out the many vulnerabilities, such as insufficient diversification of supply, political instability in supplier countries, and insufficient knowledge of the current and future outlook of material usage. The report put forward a series of policy suggestions, including propositions meant to tackle bottlenecks and price volatility through long-term contracts, stockpiles, and international agreements.

The shock of the rare-earth crisis made these vulnerabilities particularly apparent. These minerals were critical inputs for the realization of Europe's energy-technology policy launched in 2008. The SET-plan (Strategic Energy Technology Plan), as it is known, was designed to help accelerate knowledge development and technology transfer and uptake, maintain EU industrial leadership on low-carbon energy technologies, foster science for transforming energy technologies to achieve the 2020 Energy and Climate Change goals, and contribute to the worldwide transition to a low-carbon economy by 2050.[72]

The SET-Plan is meant to work in tandem with the European Industrial Initiatives (EIIs)[73] to quickly develop key energy technologies at a European level. The European Energy Research Alliance (EERA) has sought to align R & D activities across Europe to the priorities of the SET-Plan and to establish a joint-programming framework across the continent. The SET-Plan, with an estimated budget of approximately €71.5 billion,[74] provides the framework from which Europe's 2030 climate and energy policy goals are to be realized.[75] These goals reflect the EU's intention to reduce greenhouse-gas emissions by 80 to 95 percent below 1990 levels by 2050.

With so much at stake, the EU first responded to the rare-earth crisis through the EU Raw Materials Initiative in 2010. The report was intended to help industry prepare for all eventualities by identifying the most critical materials having a high economic importance, yet also increasingly facing potential supply risks. Rare earths were prominent on the list.[76] This was not the only initiative, however. It was coupled with the launch of a "rare earths diplomatic offensive"[77] to restore access to these valuable minerals. Germany was the most vocal of European countries, speaking up repeatedly about the shortage of these critical elements. Germany raised the issue at the G20 talks in October 2010.[78] Although not publicly released, a letter to the G20 was written by a broad coalition of businesses from North America, Europe and Asia underlining the repercussions of potential rare-earth shortages and asking that China not impose further restrictions on their export.[79] In fall 2010, both the EU and the WTO had stated that they were addressing Germany's concern over Chinese restrictions on rare-earth exports. The EU, furthermore, had that same year weighed the possibility of taking legal action against China's policy. In the end, however, it chose to join the United States and Japan in filing the joint complaint with the WTO in 2012.

Andrea Maresi, the press officer to Antonio Tajani, the EU industry commissioner at the time, reported that Europe had begun to stockpile rare earths, in order "to better profit from the material that we have in the EU." Maresi added, "We are trying to improve our sourcing and reduce our dependence on China."[80] Later, in November 2011, he stated that "European companies need to have a secure, affordable and undistorted access to raw materials. This is essential for industrial competitiveness, innovation and jobs in Europe. Today's report (i.e.: on critical raw materials 2010) highlights that we are on the right track with our raw materials strategy."[81]

In fact, following the commission's report on critical materials in 2010, the Joint Research Center[82] scientists found that five metals essential for manufacturing low-carbon technologies were at risk of serious shortages. The JRC scientists produced a list of recommended actions that would allow the SET-Plan to move forward smoothly to deploy and develop low-carbon applications.[83] The metals identified were neodymium, dysprosium, indium, tellurium, and gallium. The recommendations in the JRC report included reuse, recycling, and, whenever possible, substitution with less critical materials, alternative technologies, and increasing Europe's primary production by opening new or dormant mines.[84]

In October 2010, Germany's economy minister, Rainer Bruederle, reiterated, "The most important domestic source of raw materials is more recycling. We need to utilize the valuable potential of our own residual waste."[85] At present, Germany has the highest commitment to a low-carbon economy of the EU industrial nations. Germany imports raw materials worth about 80 billion euros each year. Recycling is being widely talked about and discussed, especially among industrial nations that are wary of rare-earth shortages. But such a tactic faces many challenges, both technological and in terms of cost-efficiency. If prices are high, then recycling becomes a financially attractive alternative. If prices are low, the cost of recycling is prohibitive. Nonetheless, recycling figures remain very low, and that is why the issue continues to be a top EU priority. Germany, for example, recycles only 1 percent of its natural resources. If that percentage were to rise to 10 percent in the next five years "that would be a very good achievement," said Harald Elsner, a senior geologist at the Federal Institute for Geosciences and Natural Resources.[86]

Taking a further proactive approach to the crisis, the German government signed an agreement with Kazakhstan, on February 7, 2012, to form a "partnership in the raw materials, industrial and technological spheres," focusing on rare earths.[87] The drive to secure uninterrupted access to rare earths in response to price hikes has led countries like Germany to look for opportunities to invest in prospecting, in the acquisition of mining rights, in the construction of processing plants and to extend credit guarantees. These agreements for "priority access" were accompanied by technology-transfer arrangements that are vital for developing countries like Kazakhstan. This particular agreement, furthermore, represented a public-private partnership endeavor for the Germans. When industry initiated the negotiations with twelve of Germany's largest industrial concerns, forming the Alliance for Raw Material Supply Security, the government stepped in to assist. Economic relations between the two countries have been growing in recent years. Today, Kazakhstan is Germany's third largest crude-oil supplier. Agreements totaling approximately $4 billion came after Germany had also signed an agreement with Mongolia, another nation with untapped rare-earth reserves. Clearly, Germany has taken a leadership role within the EU to avoid China's restrictions hampering its own economic and industrial objectives.[88]

As Manuel Barroso pointed out, in 2010, when he described how Europe was planning to proceed to fulfill the Union's strategic objectives in the aftermath of the economic crisis plaguing its members, "The crisis is a wake-up call, the moment where we recognize that "business as usual" would consign us to a gradual decline, to the second rank of the new global order. This is Europe's moment of truth. It is the time to be bold and ambitious."[89] The emphasis on mineral diplomacy, efficiency in production, and waste-management policies and initiatives reflected the new thinking that has entered the equation since the rare earths crisis.

Sustainability is a key goal in Europe's vision in the Anthropocene, and it means more than just diversifying its energy mix. It also marks a transition to an innovative knowledge-based, low-carbon economy that supports the EU's growth strategy. It means job creation heralded by politicians as a win-win proposition for the Union. "Solar, wind and biomass technologies have progressed most rapidly . . . Europe's renewable energy sector added 320,000 jobs between 2005 and 2009 . . . In all, the employment potential from developing the renewable energy sector is estimated at three million jobs by 2020 [and] . . . in Germany alone, employment in the renewable energy sector is forecast to rise from 400,000 today to 600,000 by 2020,"[90] said Connie Hedegaard (former EU Commissioner) in 2012. According to the 2016 renewable energy and jobs review published by International Renewable Energy Agency (IRENA), the total number of green jobs across the EU was estimated at 1.17 million in 2014.[91] According to the same review, the wind industry accounted for the majority of these jobs, with Germany, the United Kingdom, Denmark, Sweden, Greece, and Austria making the most significant contributions in this area. Uninterrupted access to critical minerals is, therefore, essential if the EU is to continue to achieve these goals.

The rare-earth crisis, therefore, prompted a long overdue examination of how to weather supply disruptions of valuable materials for its economy, especially given its transition to a low-carbon future. The 2014 EU Report on Critical Raw Materials indicated that China was in fact the most influential country in terms of the supply of the twenty most critical minerals. Included on this list of imports from China are antimony (87%), coking coal (51%), fluorspar (56%), gallium (69%), germanium (59%), indium (58%), magnesite (69%), magnesium (86%), natural graphite (69%), phosphate rock (38%), (heavy) REEs (87%), silicon metal (56%), and tungsten (85%).

The continent's resource poverty in the past was offset by expansion and colonialism, which ensured the acquisition of vital materials. Today, Europe relies on international trade. The EU's initial consensual style of looking for the most effective means of addressing the rare earths crisis has not been strong enough, however, because it has left more practical matters and new exploratory projects to its member states. Seemingly, the prevailing attitude is that industry is largely responsible for ensuring access to raw materials and for finding solutions to supply-chain vulnerabilities. In fact, the most recent Report on Critical Raw Materials for the EU,[92] gives evidence that the few national strategies of member states that were included, in large part, reflected this notion of industry and individual member state responsibility.

France (2010), Finland (2010), Germany (2010), the Netherlands (2011), the United Kingdom (2012), and Sweden (2010) were among the countries that produced reports on their raw material strategies. Sweden and Finland, for example, aimed at strengthening their mining positions and incorporating sustainability into the management of their resources. The others focused mainly on resource risk strategies; the United Kingdom emphasized "a framework for business action to address resource risk," something that Germany also underscored. The EU has mainly focused on providing the regulatory framework to help avoid bottlenecks, level the playing field, and help to maintain fair market conditions.

Throughout this period, Europe tried to maintain a balance in its relationship with China. The PRC provides a strong export market for Europe's goods, invests in Europe, and holds significant amounts of European debt. Unlike the United States, as we shall soon see, though Europe desires to maintain a powerful say on the global playing field, engaging actively in how the world works, it is not positioning itself as the rival power to China. Furthermore, in the case of the EU, the question of access to rare earths has not been so strongly linked to the vulnerabilities in the defense industry; instead, the emphasis has been on high-tech and renewable applications overall.[93] Nonetheless, if the crisis once again becomes as intense as it was in 2010 and 2011, Europe will not have set up an effective centralized plan that is sufficient to disentangle itself and its industries from China's powerful hold on rare earths.

When it became clear, in 2009, that the supply or rare earths would begin to be problematic, Japan, a major importer of rare earths, began to take steps to prevent shortages in their industries. Yukio Edano, Japan's trade minister in 2012 said, "It is important that the consuming countries and supplying countries . . . develop a global supply chain so that we are not dependent on one source, . . . If we cannot access these resources, it will slow the transition to renewables. This is not acceptable."[94] While Japan's imports from China were particularly high it was believed—and strongly alleged by China—that perhaps up to a fifth of the REEs entering Japan came from a black market network[95] that had been thriving in China.

The Japanese government was thus forced into action because of the resource crunch and complicated geopolitical tensions. As a highly industrialized economy, Japan's different product supply chains are spread all over the world. This fragmentation, characteristic of the global industrial economy, in combination with its own resource poverty and an increasingly tense geopolitical environment, put Japan in an extremely vulnerable position. This is why, even though Japan's industries are expected to address and manage supply disruptions and ensure uninterrupted access to vital inputs, the state also takes an active role by working with businesses to help offset adverse impacts on the supply chain from regional rivalries.

Japan's REE industry is intrinsically linked to China. Forty percent of China's REE exports go to Japan, in comparison with 18 percent to the United States. Japan, which boasts an exceptional high-tech industrial base, uses REEs in polishing (20%), metal alloys (18%), magnets (14%), and catalysts (12%). In 2011, 82% of its REEs originated from China.[96]

During the escalation of the rare earths crisis, Japan's strategy included increased support for mining development in foreign countries and infrastructure development in surrounding areas. Japan also promoted active technology transfer and environmental conservation. Although Japan has been energetic in the area of urban recycling of metals from compact electronics, such as cell phones and digital cameras because of their significant rare-earth content, it understands the limited viability of this effort when rare-earth prices are low. "Recycling can't be implemented immediately as it takes time for it to be a viable business. But there is no doubt it has to be done," said Naohiro Niimura, a partner at Tokyo-based research and consulting firm Market Risk Advisory Co.[97]

Given Japan's acute supply vulnerabilities, in February 2004 the government integrated two agencies, Japan National Oil Corporation (JNOC) and Metallic Minerals Exploration Financing Agency of Japan (MMAJ) to establish the Japan Oil, Gas and Metals National Corporation (JOGMEC). It assumed its predecessors' respective roles as independent administrative institutions in order to secure a stable supply of oil and natural gas and of nonferrous metal and mineral resources. JOGMEC manages Japan's stockpiles of petroleum and liquid petroleum gas. It also manages the country's national stockpiles of rare metals to ensure stable economic conditions. In addition, JOGMEC periodically reviews Japan's commodities in reserve and stockpiles rare metals to ensure stable economic conditions; rare-earth elements have been designated as meriting close observation.[98]

JOGMEC has agreed to partner with India to more actively explore for new rare-earth resources and to establish a processing facility. This includes exploration of seabed minerals, which constitutes a new high-tech frontier for mining. The model by which Japan pursues these kinds of partnerships is by backing the endeavors of Japanese firms in these regions. India and Japan have embarked on a larger strategic partnership, in addition to their September 2014 agreement, on the commercial contract between Indian Rare Earths Limited and Toyota Tsusho Corporation for the exploration and production of rare earths.[99] Japanese firms are particularly interested in the raw stage of rare earths, so they seek mining projects outside China in which to invest. Examples of such projects include the collaboration between Sumitomo Corporation and Kazakhstan's National Mining Company—Kazatomprom—which have formed a joint venture to produce LREEs. Toyota Tsusho and Sojitz partnered with Vietnam's Dong Pao project to produce LREEs.

JOGMEC also invested in the Lynas Corporation of Australia[100] in 2011. Specifically, Sojitz Corporation and JOGMEC provided a total of US$250 million[101]—through loans and equity—to receive over a period of ten years 8,500 tons of rare-earth products for Japan. The Foreign Investment Review Board in Australia approved the investment in April of 2011. Japan's proactive investment has helped to keep Lynas operational.

The mine, one of only two significant rare-earth mines outside of China, has faced many challenges in trying to stay afloat since the rare-earth crisis. Its counterpart in the United States, Molycorp, has already gone bankrupt.

Japan clearly recognizes the geostrategic implications the rare-earth crisis signifies given its growing and increasingly heated rivalry with China especially because of how it has itself been impacted. This is why its strategy reflects a sense of urgency as it seeks concrete and practical ways of weathering another possible crisis and shortage. Relying on imports, Japan's primary focus is to ensure that it does not become dependent on one supplier, and especially not China. Its emphasis, as we have seen, has been on research, recycling, and substitution as responses to the problem, in addition to seeking investments in other viable foreign projects. These endeavors have produced some positive results for Japan. Since the rare-earth crisis, Japan has been able to somewhat increase its rare-earth imports from other sources, but its imports of rare earths from China have also increased. According to figures from the Japanese Finance Ministry, (see Figure 4.4), in 2008 Japan imported 31,097 tons of rare earths from China representing approximately 91 percent of its imports. In 2010, it imported 23,311 tons, which represented 81.6 percent of its imports. By 2014, Chinese imports stood at 13,303 tons, representing 59.6 percent of its imports. Still, the imports from China were significantly up from 9,084 tons in 2013. Increases in supply from elsewhere helped offset Japan's dependency, but the 2014 increase of rare earths coming from China indicates that though the overall percentage may be lower, imports from China are again growing, as are the total imports of rare earths into Japan. Earlier, low overall imports of rare earths may have been attributable to the recession, the illegal feed of rare earths coming into Japan, stockpiling during the crisis, and also technological improvements in substitution and efficiency. The data also indicates that breaking away from China's stranglehold will not be an easy task in the future.[102]

In the United States reactions to the perceived rare-earth crisis took place at many levels, but mainly within the context of the growing geopolitical rivalry between the two powers and their economies. The climate at the time was riddled with alarmist and bold statements over Chinese intentions. According to Clyde Prestowitz, former US trade negotiator,

> The mantra in the US ever since the late 1990s has been that global-ization will make everybody rich. By being rich, they will all become democratic. By being democratic, they will all be peaceful. Well, glo-balization is working in a somewhat different way. China is getting rich and India is getting rich. But China's not getting democratic. We've seen in the recent case of China embargoing the export of rare earths that it's a kind of a mercantilist economy. The economy is being run for strategic purposes in ways that we didn't anticipate.[103]

Similarly, then US Representative Ed Markey stated, "I am troubled by this recent turn of events and concerned that the world's reliance on Chinese rare-earth materials, in combination with China's apparent

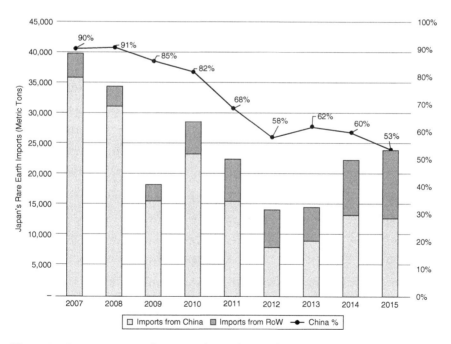

Figure 4.4 Japanese rare earth imports from China and worldwide
SOURCE: Trade Statistics of Japan, Ministry of Finance

willingness to use this reliance for leverage in wider international affairs, poses a potential threat to American economic and national security interests."[104]

Nonetheless, during the rare-earth crisis, the United States, like the EU and Japan, kept its responses within a limited range of options. US disquiet was first expressed at a high level in 2010 during Secretary Clinton's seven-nation tour of the Asia-Pacific region in November 2010. Clinton discussed the United States' growing apprehension over Chinese restrictions of rare-earth exports with her Australian counterpart Kevin Rudd and also with Japanese foreign minister Seiji Maehara. She described the rare-earth crisis as a "wake up call" for the United States and its allies to diversify their sourcing. [105]

The United States joined the other industrial nations in seeking solutions to resource scarcity, emphasizing the need for research, development, and education in rare earths to help facilitate investment in domestic production facilities and to promote international collaboration in the field.[106] The difference between Europe, Japan, and the United States, however, was that the United States has considerable rare-earth reserves in its territory and could therefore underscore the possibilities of developing its own resources instead of actively seeking mining opportunities abroad.

Moreover, and in line with the United States' tradition of innovation, researchers in private and public industries have been pursuing the development of alternative solutions.[107] In spring 2010, the US Government Accountability Office[108] submitted a report to the Committees of the Armed Services and the Senate and House of Representatives on rare earths used in the defense industry and its supply chain. The fact that rare earths became associated with matters of defense added a level of urgency to dealing with the crisis and piqued the interest of legislators in Washington.[109] Through the report, it became abundantly clear that rare-earth materials must go through a number of processing stages before they can be used in an application. Because the US Department of Defense uses rare earths in critical defense systems, the report mapped out the steps needed to mine and produce rare earths, showing that the process is long and arduous, time consuming, and capital intensive. It became evident that the production of rare earths could not effectively be established overnight.[110]

Following the flurry of reports and the alarm over resource scarcity and its impact on defense and security, a number of legislative bills were introduced in Congress. Congressional gridlock, however, ensured that none of them were ever voted into law. On March 13, 2012, Senator Lisa Murkowski (R-AK), in her keynote address at the Technology and Rare Earth Metals (TREM) Center's 12th Annual Conference, highlighted both the urgency of the situation, congressional efforts, and the frustration that she, too, had experienced having introduced the Critical Minerals Policy Act.

> Minerals are the building blocks of our nation's economy. From rare earths to molybdenum, we rely on minerals for everything from the smallest computer chips to the tallest skyscrapers. Minerals make it possible for us to innovate and invent—and in the process they shape our daily lives, our standard of living, and our ability to prosper . . .
>
> There is no question that a stable and affordable supply of minerals is critical to America's future competitiveness. And yet— despite that—our mineral-related capabilities have been slipping for decades. Rare earths garner most of the headlines, but we are 100 percent dependent on foreign sources for 17 other minerals and more than 50 percent dependent on foreign sources for some 25 more. For years, the government has been content to report on those facts—without doing much to change them.[111]

Murkowski, along with nineteen bipartisan cosponsors, had introduced the Critical Minerals Policy Act (S. 1113) in 2011. According to Murkowski's statements during the conference, "The Senate Energy and Natural Resources Committee still has not allowed the bill to be marked up, in part due to misperceptions about its permitting and resource assessment provisions."

The Senator went on to say that while the problem had drawn the attention of legislators, none of the bills introduced at the time had even come close to becoming the law of the land. "In the Senate alone, twenty-four different Senators—nearly a quarter of those serving—are supporting legislation to address some aspect of this problem . . . So far, those efforts have fallen victim to the new normal in Congress: they've gone nowhere. Not one bill on this topic has been reported from a Senate committee—even when the votes are likely there to do so." The following is a list of legislative initiatives that drew attention but did not become law, highlighting the inconsistency between declared objectives and the realities on the ground.

- On March 17, 2010, the RESTART Act (H.R. 4866) was introduced by Representative Mike Coffman (R-CO) regarding the stockpiling of rare earths and the establishment of rare earths production facilities in the United States. According to the 2011 Department of Energy report on critical materials,[112] the goal of this bill was to "reestablish a competitive domestic RE minerals production industry; a domestic RE processing, refining, purification and metals production industry; a domestic RE metals alloying industry and a domestic RE-based magnet production industry and supply chain in the Defense Logistics Agency of DOD." No further legislative action has been taken since November 28, 2011. In effect, this bill, too, "died" by being referred to committee.[113]
- In the Senate, the Rare Earths Supply Technology and Resources Transformation Act of 2010 (S.3521) was introduced by Senator Murkowski. The Senator herself pointed out that this piece of legislature has gone nowhere.[114]
- In September 2010, the House of Representatives passed H.R. 6160, a bill that directed the Department of Energy to support new rare-earth technology through public- and private-sector collaboration and coordination with the European Union. The same bill called for loan guarantees for rare-earth-related investment. The bill was introduced to the House by Representative Kathleen Dahlkemper (D-PA). The bill died because it was never passed by the Senate.[115]

- On March 8, 2011, Representative Brad Miller (D-NC) introduced
  H.R.952, Energy Critical Elements Renewal Act of 2011, to
  develop a rare earths material program, to amend the National
  Materials and Mineral Policy, Research and Development Act of
  1980, and other purposes.[116] The bill was never enacted.

The rare-earth crisis may have died out in the news, but individual members of Congress keep trying to draw attention to the inability of the US to devise a long-term solution. From the point of view of the legislative process, nothing has come of these efforts except an indication that the particular problem has not been resolved satisfactorily. In 2014, Steve Stockman of the House of Representatives introduced Bill HR 4883 that sought to establish a "Thorium-Bearing Rare Earth Refinery Cooperative as a federal charter to provide for the domestic processing of thorium-bearing rare-earth concentrates as residual unprocessed and unrefined ores." The bill further required the "Cooperative's Board to establish a refinery and a Thorium Storage, Energy, and Industrial Products Corporation to develop uses and markets for thorium, including energy." It also "directed the Secretary of Defense (DOD) to coordinate with other federal agencies to advance and protect domestic rare-earth mining, the refining of rare-earth elements, basic rare-earth metals production, and the development and commercialization of thorium." The bill went on to mandate that "beginning in January 2020, all purchased or procured weapon systems to contain only U.S. or North Atlantic Treaty Organization (NATO) member nation produced and sourced rare-earth materials, metals, magnets, parts, and components." It also called for the prohibition of "any rare earth materials that originate or pass through a non-NATO member nation and barred any waivers from being granted unless the lead contractor can demonstrate that it has pursued all possible corrective actions, including direct investment into the supply chain."[117] The bill would need to pass by both the House and Senate in identical form to then be signed by the president to become law. It was never enacted.

The list of legislative initiatives included H.R.761 National Strategic Minerals Production Act of 2013 that was received, read twice, and referred to the Committee on Energy and Natural Resources.[118] It too was never passed by the Senate. H.R 1937: National Strategic and Critical Minerals Production Act of 2015 passed the House on October 22, 2015, but was never passed by the Senate.[119] Moreover, the White House issued a statement in which it expressed the administration's opposition to the bill arguing that while it "strongly supports the development of rare earth elements and other critical minerals, [it] rejects the notion that their development is incompatible with existing safeguards regarding uses of public lands, environmental protection, and public involvement in agency decision-making."[120] In summary, repeated attempts at legislative action to respond to United States' dependence on rare-earth imports from China have yet to produce any concrete results.

In searching for the appropriate domestic response to the crisis, the Department of Energy put together a report for the development of a Critical Materials Strategy in 2010. The main objectives were threefold. First, to mitigate supply risk, the United States would need to diversify its global supply chains. Second, it would have to develop both material and technology substitutes. Third, it would need to promote recycling, reuse, and efficiency of use in order to lower dependence on critical materials. Although the report included other critical materials as well, it was prompted by the rare-earth crisis. The Department of Energy updated its report, publishing its Critical Materials Strategy in December 2011.[121] In the strategy itself, it became clear that the department was particularly concerned about how rare-earth shortages could impact clean-energy technologies. There was acknowledgment throughout the report that while demand for critical materials had increased over a decade, there were factors that had prevented the supply from catching up. Capital constraints, long lead times, trade policies, the complexities of coproduction and byproduction, as well as the market's lack of transparency and size, were contributing factors to its lack of efficiency.

In any event, and as David Sandolow, assistant secretary for policy and international affairs at the Department of Energy, underscored, on March 17, 2010, at the Technology and Rare Earth Metals Conference in Washington DC, "Supply constraints aren't static. Strategies for addressing shortages of strategic resources are available, if we act wisely. We can invest in additional sources of supply. We can develop substitutes. We can re-use materials and find ways to use them more efficiently. We can consider use of stockpiles and strategic reserves."[122]

Domestically, the United States once again turned its attention and focus toward research and innovation, which is why it also decided to fund an Energy Innovation Hub that could produce solutions to domestic shortages of critical materials such as rare earths should these ever occur again.[123] The Critical Materials Institute limits its focus to "research and development efforts leading to technologies that can diversify the sources of critical materials; provide substitutes for materials that are in short supply; or improve the utilization of existing resources through enhanced efficiency in manufacturing and improved recycling . . . We measure our progress by the advancement of the relevant technologies, and our success by the adoption of our technologies by the commercial sector."[124]

Correspondingly, although there was considerable hand-waving at the Department of Defense very little progress in addressing possible shortages has materialized. The United States created the first National Defense Stockpile in 1939. The stockpile was meant for use during situations of national emergency, and the goal was to maintain and manage strategic and critical materials for just such an eventuality.[125] According to the "Reconfiguration of the National Defense Stockpile (NDS) Report to Congress" (2009),[126] the Department of Defense determined that its stockpile was in excess of its needs. Congress authorized disposing of more than 99 percent of the stockpile's material and earmarked the money from the sales for various defense programs, primarily military health and retirement benefits. The stockpile did not contain rare earths, and revenue from the sales amounting to more than $5.9 billion between 1993 and 2005 have also been used as a tool to strengthen relations with countries with which the United States was seeking to build relationships.[127]

In its 2008 report, the Department of Defense defined what "criticality" means for its mission. "The "criticality" of a material is a function of its importance in DOD applications, the extent to which DOD actions are required to shape and sustain the market, and the impact and likelihood of supply disruption."[128] In a report to Congress, in April 2012, entitled "Rare Earth Elements in National Defense: Background, Oversight Issues, and Options for Congress," Valerie Bailey Grasso, a specialist in defense acquisition, proposed that "Congress could require a strategic rare earth elements stockpile. Stockpiles might possibly increase the security of the domestic U.S. supply for rare earths."[129] In its 2013 Strategic and Critical Materials Report on Stockpile Requirements, the Department of Defense recommended stockpiling $120.43 million of HREEs. The Strategic Materials Advisory Council went one step further to urge the Department of Defense to "create and nurture a U.S. based rare earth supply chain."[130] According to the US Geological Survey report of 2016, the United States did not stockpile rare earths.[131]

In February 2016, the US Government Accountability Office submitted a report to congressional committees about rare-earth materials. It found that six years after the crisis, the Department of Defense had not yet developed a "comprehensive approach for ensuring a sufficient supply of rare earths for national security needs—one that can establish criticality, assess supply risks, and identify mitigating actions—[that] would better position DOD (Department of Defense) to help ensure continued functionality in weapon system components should a disruption occur, even though supply disruptions in rare earths have not occurred over the last several years."[132] Furthermore, and according to the same report, the Department of Defense had not reached an agreement on what in fact "constitutes 'critical' rare earths. While various organizations' definitions of critical may be similar, DOD has identified 15 of the 17 rare earths as critical over the last 5 years."[133]

The Obama Administration was slow to respond as well. Two years after the crisis and just three days after announcing the decision to take China to the WTO, President Obama, acting in his capacity as Commander in Chief of the Armed Forces, went on to sign an executive order on resources preparedness. The order included a substantial section on the expansion of productive capacity and of supply, and it called for loans, loan guarantees, subsidies, and more. There was a particular focus on critical and strategic minerals.

Sec. 306. Strategic and Critical Materials. The Secretary of Defense, and the Secretary of the Interior in consultation with the Secretary of Defense as the National Defense Stockpile Manager, are each delegated the authority of the President under section 303(a)(1)(B) of the Act, 50 U.S.C. App. 2093(a)(1)(B), to encourage the exploration, development, and mining of strategic and critical materials and other materials.

Sec. 307. Substitutes. The head of each agency engaged in procurement for the national defense is delegated the authority of the President under section 303(g) of the Act, 50 U.S.C. App. 2093(g), to make provision for the development of substitutes for strategic and critical materials, critical components, critical technology items, and other resources to aid the national defense.[134]

After the initial crisis erupted, analysts tried to explain the timing of these decisions. Many attributed them to tensions with Iran that could potentially lead to an all-out conflict, and this was seen as a way for the administration to put China on notice. Others interpreted the decision as a tacit acknowledgement that the US military was in fact vulnerable to China's monopoly of rare earths, including its control of the supply chain.

That same month, the US government moved even more aggressively with two targeted decisions with economic repercussions. The United States, on March 21, 2012, ruled to add a customs tax on China's solar panels. The decision followed a long investigation into Chinese state subsidization of the solar industry, which had brought prices down by approximately 30 percent since the PRC began to move robustly into their manufacture. In 2011, the United States imported $3.1 billion in solar panels from the Chinese. The United States' taxation of Chinese solar companies would be determined by the level of subsidy thought to have been given by the PRC government. Examples of the tax levels imposed were a 2.9 percent tax on China Suntech; whereas Changzhou Tina was taxed at 4.73 percent. Other companies were taxed at 3.61 percent. The EU followed suit with more modest taxation. These consecutive actions were aimed at putting pressure on China to desist from trying to control the renewables industry in which both the EU and the United States had a growing stake.

The Chinese side, of course, did not remain silent as US reactions escalated. Chinese companies argued that the United States also offers subsidization for its companies. The Chinese government counteracted US actions by declaring that it would launch its own investigation into the United States' renewable energy practices. In retaliation, the PRC decided to tax polysilicon, the main ingredient in solar production, which the United States exports to China.[135]

These vocal and forceful initiatives, undertaken by the United States within a single month, were meant to signal two things. First, they fired a number of warning shots to encourage the Chinese government to curtail or rethink its strategic designs at not only an economic but also a geostrategic level. It also reflected a growing US interest in having a role in the production and deployment of renewables. President Obama made that plain in his State of the Union Address in 2013 when he declared, "As long as countries like China keep going all in on clean energy, so must we."[136]

Having accepted that the United States was lagging behind in the green-growth race in which China was investing heavily, the Obama administration turned its attention to deploying these technologies. In point of fact, however, as much as rare earths have figured into policy discussions and the geostrategic conversation, securing unobstructed access to these materials was largely left to the industries themselves, in the same manner that both the EU and Japan had largely done as well.

In the end, two government initiatives stood out as the most solid and targeted responses to the rare-earth crisis. The first was the trilateral cooperation between the United States, the EU, and Japan, and the second was the filing of the case against China at the WTO. Both garnered much attention and publicity, but a closer examination will reveal their limited scope and effectiveness.

## TRILATERAL COOPERATION

When Japan, the EU, and the United States joined forces to find mutually beneficial solutions to the rare-earth crisis, they were attempting to address a number of wider problems. Clean-energy options required a large quantity of rare earths and other less common materials. These materials were supplied by a potential rival and as markets continued to grow, supplies could become even tighter and costs more prohibitive. These facts, then, raised the following questions: How could the EU, Japan, and the United States find new or enhanced recycling technologies to increase the available supplies of rare earths? Is substitution possible, and are there alternate device designs that might perform as efficiently at a comparable cost? How can changes in design and technological innovation impact the amount of rare earths necessary to give the optimal result of an application?

The first workshop was organized by the European Commission, the US Department of Energy, and the Japanese Ministry of Economy Trade and Industry, as well as the New Energy and Industrial Technology Development Organization. It was held in Washington, on October 4 and 5, 2011. The main issues addressed were the policy and strategic implications of shortages, and these were followed by two parallel technical workshops. One workshop explored techniques of extraction, separation, and processing in a sustainable manner, and the other focused on efficient uses and substitutes. These themes carried into the second, third, and fourth conferences, held in Tokyo in 2012, Brussels in 2013, and at the Ames Laboratory in the United States in September 2014, respectively.

The dialogue that the trilateral workshops initiated facilitated the exchange of best practices and common approaches to potential shortages of critical metals. In the beginning, they drew considerable media and government attention, but since the end of the rare-earth crisis, they seem to have faded to the background.

## CHINA AND THE WTO

China became a member of the WTO in December 11, 2001. This was thought of as an historic event. China's population size, its rapidly expanding economy and its one-party system in which the state maintained a significant role vis-à-vis resource allocation – all made it a formidable rising international power.. The fear of other WTO members was whether or not China's economy would blend in with theirs, which were primarily market oriented. As noted by economist Robert Lawrence,[137] when China joined the organization, there were worries that it "would not participate constructively in the WTO. It would throw its weight around, try quickly to obtain disproportionate influence and use its influence to fundamentally change the WTO system. China was also seen as a potentially powerful addition to the ranks of developing countries, and many in the developed world worried that it would seek to limit the obligations required of developing countries." The logic that prevailed, however, was that it would be far better to have China engaged in the operations of the world system through its participation in international organizations than to have it remain isolated at the edges of that system, where it could threaten and challenge international order and stability as the other actors understood those to be.

For China, too, the process of accession held significance.[138] Accession had never been straightforward because there were many voices in China that opposed such an opening of the economy. Membership was a complex commitment, and China's internal debate reflected considerable hesitation. There were those who argued that China (ever since it began opening its economy to the world) had been relying rather heavily on foreign investment, and that it was now time for it to develop the national domestic economy. The contagion of the 1997 Asian financial crisis seemed to strengthen these anxieties. Nonetheless, China "needed the rules-based WTO system to secure rights to market access for exports and rights against protectionist measures of its trading partners, as it was moving to the very center of the globalization process. And the WTO needed China as a full and committed member to be a truly global and effective system."[139]

Ultimately, after internal bargaining at the highest Chinese levels, it was decided that it would be best to proceed by opening a number of economic sectors to competition and to carry out economic reform, especially in the public and banking sectors. The time had come for China to reciprocate if Chinese exports were to more substantially compete in the markets of many of its trading partners. The high-tech revolution of the 1990s was another contributing factor to China's decision because it could not afford to miss participating in these technological developments. Finally, China's entry into the WTO would allow it to partake in the workings of the organization and help shape its future without having to depend on revisions for most-favored-nation status.

For all these reasons, both sides, understandably, had their own apprehensions at the time of accession. China's inclusion could have caused disruption and an unprecedented trade surge. But mostly there was fear that China would enter the WTO and then proceed to disrupt its workings by ignoring its rules. How could the others be sure of China's commitment? The United States and the EU had the greatest concerns, but they took the risk and undertook the monitoring of China's behavior and compliance through task forces and by using two important institutional mechanisms: the Trade Policy Review Mechanism and the Dispute Settlement Body. They had also requested that prior to membership, China should first accede to a protocol under which the PRC would commit itself to wider non-WTO obligations.[140]

As Mike Moore, the WTO director general, observed at the conclusion of the meeting of the Working Party on China's Accession, which took place in Doha, "International economic cooperation has brought about this defining moment in the history of the multilateral trading system . . . With China's membership, the WTO will take a major step towards becoming a truly world organization. The near-universal acceptance of its rules-based system will serve a pivotal role in underpinning global economic cooperation."

Since China's accession to the WTO, many have written about the subsequent experience of those years and on China's performance as a member. The general consensus seems to be that China's participation has been positive for overall trade. During the first few years, the PRC moved conservatively inside the WTO system, perhaps to better understand it and get its bearings, but it gradually became more active in the workings of the organization and of the Dispute Settlement Body system as well. The Dispute Settlement Mechanism of the WTO is rules oriented. The settlement of disputes is done through a set of enforced rules previously agreed on by both parties. Given that the disputes are brought by states against other states, this system provides a number of benefits that induce acceptance and a willingness to settle.[141] China's initial reluctance to participate in the Dispute Settlement Body mechanism may have reflected a suspicion of normative constraints and even an aversion to multilateral adjudication.[142] Nonetheless, China has complied with tariff-reduction commitments in a timely fashion.[143]

Since those early years, moreover, China has increasingly made use of the mechanism for settling disputes, many of them between the PRC and the United States. This in the minds of some indicates that these initial disputes could be foreshadowing possible trade wars. Specifically, from 2007 and 2012, the United States brought thirteen WTO cases against China, and China brought seven against the United States.[144] In the end, there will always be some critics who offer a consistent critique of China's "maximizing its interests through minimal involvement abroad."[145] This critique comes mostly from those who view China as a larger threat because of its growing economic, political, and military strength.

On March 13, 2012, the United States, the EU, and Japan chose to act in unison by simultaneously filing complaints with the WTO demanding consultations with China over its restrictions on the export of rare earths, tungsten, and molybdenum.[146] The three powers alleged that China's actions were not in line with WTO provisions. The actions they listed were:

> The imposition of export duties; The imposition of export quotas, and other quantitative restrictions; The imposition of other restrictions such as the right to export based on licenses, prior export experience, minimum capital requirement, and {other conditions that appear to treat foreign invested entities differently from domestic entities}; The maintenance of minimum export prices, through the examination and approval of contracts and offered prices, and through the administration and collection of the export duties, {in a manner that is not uniform, impartial, reasonable or transparent};

The imposition and administration of restrictions through unpublished measures.[147]

European Union trade commissioner Karel De Gucht said that China's restrictions on rare earths "hurt our producers and consumers in the EU and across the world, including manufacturers of pioneering hi-tech and 'green' business applications."[148] In its response, China insisted that it was cutting rare-earth mining because of environmental concerns. "Regarding rare earth management, we have a very clear idea and direction, which is environmental protection and the long-term sustainable use of resources."[149]

The complaint against China was subject to divergent interpretations. Professor Yufan Hao of the University of Macau, in 2012, for example, during an interview for the National Bureau of Asian Research said,

> The basis for the three complainants' case is the WTO's support for free trade, and that the three parties think that China's export restrictions on rare earths are against WTO rules. However, many people in China feel that this case is quite ironic. These countries never complained to the WTO about China previously dumping underpriced rare earth product, as they did with China's export of low-priced steel and textiles. They urged China to sell them rare earths at a very low price, and denounced rare earth export restrictions from a liberal economic viewpoint. And at the same time, these three groups are also very reluctant to sell China high-tech products, not to mention arms, which are produced using rare earths.[150]

Others raised questions about China's defense strategy. Jane Nakano, a fellow in the Energy and National Security Program at the Center for Strategic and International Studies, had the following concerns:

> China's dominance of the global rare earth supply has come at a great cost, with serious environmental issues. But many consumer countries feel that China will have to provide a much more satisfactory

answer as to why the export quota has been declining while the production quota has been increasing. Also, it's one thing to have the overall level of export quotas unchanged, but it would be quite another to allow exports—in a sufficient amount—of the types of rare earth materials that consumers want.[151]

The EU, the United States, and Japan took their complaint to the WTO in accordance with their wider policies of seeking to find solutions with China through an institutionalized, cooperative, rules-based framework to which all parties have subscribed. It was also their way of not merely settling a trade dispute but also using a tool of economic statecraft to challenge the Chinese position.

This became particularly clear when President Obama himself stepped into the Rose Garden to announce to the world that the three allies had taken action. Obama, with much fanfare, was responding to China, not just to a material shortage of rare earths. His announcement was tailored to both an international and a domestic audience. As the US president faced criticism on an economy still in recession, Obama's words were greeted partially as a predictable political maneuver to emphasize the government's concern with job creation in the United States. In the Rose Garden and under the guise of the rare earths dispute, the president could defend American interests, promote renewables as a response to climate change, and put China on notice, all in one fell swoop. On an international level, he indicated that the United States would continue to defend fair-trade practices, protect its dominant position both in the high-tech and in the green-tech industries, and champion a system of international rules and norms. Accordingly, President Obama underlined that American manufacturers needed unobstructed access to rare earths in order to produce high-tech products such as advanced batteries and that by curtailing exports, China was not allowing them to do so. This, he stated, went against WTO regulations. "Being able to manufacture advanced batteries and hybrid cars in America is too important for us to stand by and do nothing. We've got to take control of our energy future, and we can't let that energy industry take root in some other country because they were allowed to break the rules . . . We are going to make sure that this isn't a country that's just known for what we consume."[152]

The president, the most vocal of all three actors in the dispute, strove to ensure that the clean-energy agenda that his administration had been pushing would have the materials necessary for production. Reliable and affordable access to rare earths was essential to achieving this aim. The steep rise in prices following the 2010 crisis and the fear that exports could be further restricted in the future raised concerns in green-technology companies in the United States because rare earths were essential elements in their industry.

It had not been the first time China had interfered with the supply of critical materials. On the contrary, it had restricted raw-material exports before. In 2009 the United States, the EU, and Mexico launched a WTO case challenging China's right to restrict bauxite, coke, magnesium, manganese, and zinc exports, again forcing prices to rise.[153] On January 30, 2012, a WTO panel ruled that China was in fact in violation of WTO regulations.[154] The panel asserted that the restrictions led to price inflation outside China and gave domestic Chinese firms unfair advantage. The ruling encouraged the United States, Japan, and the EU to submit a similar case about rare earths on March 13, 2012.

Immediately following the ruling on bauxite, coke, magnesium, manganese, and zinc, Vivian Pang, an analyst with the Asian Metal consultancy in Beijing, said, "It is still too early to say what the impact will be [on rare earths], but I can't see it having a big impact on prices—the main issue will still be supply and demand."[155] According to industry analysts at the time, the widely held belief was that China would most likely not increase its production capacity because its priority was to control the environmental repercussions due to lax oversight in mining and separation and in order to maintain high prices of their strategic elements. The use of the environmental defense at the WTO was, in fact, grounds for limiting exports, so China needed to demonstrate the health and environmental implications of REE mining, and to show how the reduction of production had in fact helped to cut pollution and improve public health. China had also to convince the WTO that it applies its policies equally to foreign and domestic companies.

According to Dr. Si Jinsong,[156] at the time the second secretary of the Economics Affairs Office at the Embassy of China in Washington DC, who spoke at the TREM 12 Conference on March 14, 2012, China would continue to supply the global market with rare earths. It would also develop its policies and effectively manage its resources in line with WTO regulations. The PRC, according to Dr. Jinsong, would strengthen its international cooperation with the United States, in particular, especially in areas of substitution and the improved utilization of these resources. He insisted that China had been urging other countries to develop their own rare-earth resources—since they do exist elsewhere—instead of only turning to China for supplies.

Dr. Jinsong spoke at length about the need to consolidate the industry within China, which once had 1,000 REE mining, smelting, and separating enterprises, but at the time had only 120. He attributed this reduction to the need to monitor the environmental toll of extraction and processing in many parts of the country, as well as the Chinese decision to crack down on the illegal mining and smuggling[157] of rare earths.

Dr. Jinsong produced extensive data highlighting that thirty-four countries had discovered rare earths, adding that China's rare earths accounted for 36.4 percent of the global total. He stressed that China's per capita reserve of rare earths was lower than the reserves of other countries, emphasizing that mining had led to China's reserves quickly dwindling, falling by 40 percent in the past fifty years. China, he continued, had been exporting beyond its reserve share, and it now had to be careful because if the reserve was exhausted, it would be damaging not only to China but also to the world economy. In short, China's position was that the world had received ample warning that it needed to look elsewhere for rare earths, because if China remained the only reliable supplier, then it would be safe to surmise that such a trajectory would lead to conditions of scarcity because of limited market availability.

To underscore these concerns, the Chinese Information Office of the State Council released a white paper on rare earths in 2012. It was drafted to "provide the international community a better understanding of this issue."[158] According to its findings, excessive mining taking place over the past fifty years had led to such rates of decline that in rich rare-earth areas like Baotou,

> only one-third of the original volume of rare earth resources is available in the main mining areas, and the reserve-extraction ratio of ion-absorption rare earth mines in China's southern provinces has declined from 50 two decades ago to the present 15. Most of the southern ion-absorption rare earth deposits are located in remote mountainous areas. There are so many mines scattering over a large area that it is difficult and costly to monitor their operation. As a result, illegal mining has severely depleted local resources, and mines rich in reserves and easy to exploit are favored over the others, resulting in a low recovery rate of the rare earth resources. Less that 50 percent of such resources are recovered in ion-absorption rare earth mines in southern China, and only ten percent of the Baotou reserves are dressed and utilized.[159]

Nevertheless, the Chinese arguments did not suffice, and the PRC did not win the case at the WTO. The ruling came out against China on March 26, 2014. China, moreover, lost its appeal, on August 7, 2014, and subsequently announced its compliance, which took place in May 2015. While the dispute lasted, there had been much speculation about what China would be willing to do vis-à-vis rare earths. Would it abide by its WTO obligations or not? The outcome of the dispute and China's compliance indicates that the PRC—just like other members—had gained the expertise on how to adhere to the rules and regulations of the organization while simultaneously using the rules to promote and secure its own interests.

A recent study published in *Resources Policy*,[160] in 2015, examining whether or not China's rare earths export policies had worked over the period in question showed that, in fact, "the market power and price sensitivity of China's rare-earth products increased dramatically, indicating that China's export policies have exerted significant effects. The quantitative estimate of the market power of China's rare-earth products on the US and Japanese markets shows that it has risen significantly." However, the authors also suggest that in the future, and in terms of sustainable improvement in pricing power, China's focus could shift from "controlling exports to controlling production."[161]

Furthermore, by the time China was required to comply with the WTO ruling, it had nothing to lose. Prices had dropped, and there was an over-supply in the market because of ongoing smuggling and the stockpiling at the height of the crisis. While the case was being discussed, China had taken advantage of the critical years it needed to consolidate the indus-try domestically in order to reposition itself for the future. As the famous Chinese strategist Sun Tzu said, "Ultimate excellence lies in not winning every battle but in defeating the enemy without ever fighting."[162]

# Conclusion

## *Rare Earths: Paradigms of Connection and Disruption*

World politics are shaped by defining events. Such events need not only be military conflicts, as were the two world wars in the last century or the present ongoing war on terror. They can also be economically driven, especially in a world as globalized as the one we are living in today. Already, the growing economic importance of the BRICS and China, in particular, has impacted the priorities, outlook, practices, and power dynamics of international institutions and world affairs.

In 2010, the rare-earth crisis that erupted over China's decision to essentially halve exports produced a ripple effect throughout the world. It highlighted prospects of intense resource competition, raised economic and technological concerns throughout industry, underlined security issues for militaries, and ushered in debates about whether conflict or cooperation would pave the way to a resolution. Reactions from countries affected by China's policy change were at the time mixed, certainly piecemeal, and neither highly strategic nor effective. Accordingly, the rare-earth crisis offers a singular case study from a variety of perspectives.

China, a behemoth of a country, exerts a near monopolistic control over these strategic elements as it continues to dominate the global supply of REEs. With a clearly calculated effort to buy into and control any rare-earth projects beyond its borders, the PRC is a powerful nation-state and an economic giant vying for a strategic role in the high-tech race.

By the same token, it is rapidly preparing for a post-fossil-fuel world by leading in the production and use of renewable energy resources. As a result, not only does the international community have to face a country that monopolizes strategic material—above and beyond the geographic singularity that oil has enjoyed over many decades—but one that also controls the technology, mining, metallurgy techniques, and the entire supply chain associated with rare earths and their applications.

Nonetheless, as important as rare earths are as a resource problem, in and of themselves it is not likely that they constitute enough of a catalyst to provoke armed conflict. However, underlying resource competition are various other independent variables at play that can affect policies, events, and the final outcome of this challenge. China's rise, growth, increasing defense spending, and its willingness to use its rare earths monopoly not only to further strengthen its economy, but also for geopolitical ends, are some of those factors that are worrisome to the world system.

The shift to a multipolar world following the fall of the Soviet Union has prompted much speculation in an attempt to anticipate and understand China's emerging global role. Will it lead, will it follow, will it remain inward looking and largely noncommittal to past paradigms and expectations of super power politics? Geopolitical tensions have already reared their heads. The United States has shifted its strategic attention toward the Asia-Pacific region and even though the new Trump administration may not be calling it a "pivot," a climate of suspicion and rivalry remains pervasive. In China, Chairman Xi is proving to be a dynamic diplomat with a globally appealing economic and trade agenda of his own. The die has been cast, and already the world is impacted by the emergence of a new set of realities in international relations structured by the growing rivalry between the United States and China.

The US decision to shift its attention from the Middle East to Asia, having first opened Pandora's box in the former, has left Europe alone to deal with a wave of mass migration of refugees and increased terrorist attacks on its soil. Japan,[1] however reluctantly, has also been forced to re-evaluate its strong pacifist stance following World War II.[2] The crisis over the disputed islands in the East China Sea will inevitably escalate over time. These geopolitical shifts should not be underestimated. They raise red flags and require an overall new assessment of the situation on the ground. As Choucri and North argue, "A crisis is only a small tip of a submerged iceberg of competitions, antagonisms, relatively nonviolent conflicts, arms races, and previous crises . . . there are [also] submerged phenomena such as population differences, technological growth, differential access to (and competition for) resources, trade, markets, and influence, the expansion of national interests, and so forth,"[3] influencing the path to conflict or cooperation. It seems safe, therefore, to assume that international conflict cannot be attributed to a single cause. A critical set of variables come into play that can determine final outcomes, including and, perhaps ultimately, leaders' perceptions of these variables themselves.

Practitioners and scholars have looked to theories of international relations as a way of understanding looming global challenges and ways to meet them. A number of different frameworks for analyzing the causes of war and the conditions for peace, especially following the two traumatic world wars in the first half of the twentieth century, emerged as tools to make sense of a quickly changing terrain. The attempt to explain power politics in the post–Cold War world has engendered others. Globalization, the clash of civilizations, the end of the nation-state, the proliferation of nonstate actors and agencies, political ecology, resource competition, economic statecraft, greed and grievance—to name a few—are frameworks that seek to predict the trajectory of world affairs and to define the international landscape.

As China takes its place on the international stage, its particular model of state capitalism and of centralized economic and strategic planning has allowed it to exploit the vulnerabilities of mutually competing democracies for its own strategic benefit. Rare earths provide one of the most salient manifestations of what forces may be shaping this new political, economic, and security environment. Although democratic governments in responding to China's handling of the rare earths crisis achieved a certain level of cooperation (mostly scientific and diplomatic), they did not formulate or enact a central strategic plan that would over time provide a viable alternative to China's dominant position in the industry.

Perhaps their weakness arises also from the fact that they are unwilling or unable to control industrial actors having knee-jerk reactions to supply disruptions. They have been unwilling to intervene in a way that would prohibit industry from moving operations into China or from abandoning non-Chinese producers of rare earths as soon as the prices of Chinese rare earths were once again lower than their competitors. Although these democracies have resorted to diplomatic overtures, workshops, and the filing of a case at the WTO to show that, in their estimation, this was more than merely a trade dispute, in time, they acquiesced to the belief that the curtailment of their access to these elements would no longer pose a threat. Instead of hanging a lantern on the problem, they have opted to underestimate the ramifications of a rivalry that they have helped to create, robbing themselves of the opportunity to clearly assess their weaknesses and to better prepare for future disruptions.

Democratic governments have been disposed to encourage the investment by businesses in new rare-earth exploration and to provide technological assistance and exchange expertise in order to foster partnerships with producing countries, such as Kazakhstan and India. They have attempted to legislate and even to generously fund research in recycling and substitution. Some have opted to stockpile and study better resource management systems. They have been able to produce comprehensive reports on military vulnerabilities that loom because of potential disruptions in access to rare earths and their impact on the security environment.[4] They have not been particularly efficient, however, in designing and centrally implementing policies of direct funding through state resources. Nor have they, for instance, managed to recreate all the elements needed to achieve a level of autonomy from China—that is, the entire supply chain, which has currently moved away from countries in the EU, the United States, and Japan.

In an ironic twist, theories of globalization, reliance on the free market, and faith in international cooperation as a way of resolving such global problems may have allowed democratic countries to develop a false sense of security. Although the rapid tide of globalization undoubtedly has had an impact on international affairs and has made the world seem a smaller, more interconnected place, states and state interests have not disappeared from the equation.[5] While the global reach of the new challenges cannot be denied, this has not stopped China from adroitly using its advantage in navigating the process by making overtures whenever it deems necessary, spreading its influence where it can, and by continuing to maintain an important component of central state planning and goal setting that it follows unwaveringly. This single-minded focus allows China to prevail in the competition for resources, to corner the market for high-tech and clean-energy applications, and to continue to grow its economy more effectively than its competitors.

China's handling of salt and oil, two strategic commodities in two different eras, provides some helpful historical insights. Repeatedly and at different times, China has developed a carefully crafted, centralized plan to capitalize on a resource it identified as strategic and vital to its interests to strengthen the state and its domestic economy. It has followed the same strategy in the case of rare earths, which, in all likelihood, will remain strategic, since the innovation necessary for efficient substitution is still, at best, years away.

Are China's policies a threat to other industrialized powers, and are they a source of growing tensions in international affairs? Certainly, China's economic and political rise is a source of anxiety for those nations that have dominated world affairs during the past several decades. The United States and Europe are carefully watching China's growing influence around the world and continually engage the PRC within the international community so that it does not remain isolated and unchecked. The expressed rivalry is particularly pronounced between the United States and China as they wrangle over trade, alliances, intellectual property, cyberespionage, spheres of influence, and views of how the world works. For Japan, centuries of economic, military, and political competition with China, as well as ethnic tension in conjunction with the still recent historical events of the twentieth century, shape its movements in Asia.

In terms of international relations in the twenty-first century, China is inevitably the elephant in the room. Its sheer size—population and territorial—and its robust economic growth, coupled with dynamic diplomatic efforts at building solid, nonjudgmental relations with countries around the globe, is an unsettling trend even though in making these moves China has been careful until now not to overtly antagonize its main rivals. Though there is a fear of its power becoming an international threat, China is proceeding according to plan without showing too aggressive a hand. In the case of rare earths, the world has now gotten a first glimpse of the tip of the iceberg and what may be lying beneath the surface. China, on the one hand, has been putting the world on notice by openly stating that other nations need to look elsewhere for these valuable materials. At the same time, it knows full well that it has carefully orchestrated conditions that ensure its own continued monopoly. China will continue to monopolize the industry. Price fluctuations will not help offset the PRC's dominant position, since they will only discourage new investments, thus strengthening China's hand.

The case of rare earths, therefore, provides an important window into a new set of international challenges. The world must now contend with a major power's dominance of crucial strategic materials in a way that is unparalleled in the past. We have seen how China has come to achieve this dominance, and the world has been given an initial glimpse of its willingness to use it to further its political and economic goals. At the same time, we have seen the inability of its international competitors to remedy this asymmetry, to effectively modify China's strategy, and to prevent spillover effects in geopolitical balances. As elements, rare earths are enablers. As political instruments, they are increasingly turning into a catalyst in a new era of potentially fraught international relations.

INTRODUCTION

1. "Rare Earths Statistics and Information," U.S. Geological Survey, January 29, 2013, accessed February 10, 2013, http://minerals.usgs.gov/minerals/pubs/commodity/rare_earths/index.html.

2. The definition of "strategic" is something that will be discussed more extensively. I am using the term more broadly and additionally to mean an item for which the marginal elasticity of demand is very low and for which there is no readily available substitute.

3. OPEC, "OPEC Share of World Crude Oil Reserves," accessed September 23, 2016, http://www.opec.org/opec_web/en/data_graphs/330.htm. "According to current estimates, almost 81% of the world's proven oil reserves are located in OPEC Member Countries, with the bulk of OPEC oil reserves in the Middle East, amounting to 66% of the OPEC total."

4. The term "monopoly" in this context is used to indicate China's excessive market power in the rare-earth industry. Not only does it control a scarce strategic resource but also it deliberately erected barriers to manage its flow. See William J. Baumol and Alan S. Blinder, *Economics: Principles and Policy* (Mason, OH: South-Western Cengage Learning, 2011), 217–62.

5. "Critical Raw Materials," European Commission, "Growth," accessed September 23, 2016, http://ec.europa.eu/growth/sectors/raw-materials/specific-interest/critical/; "Tackling the Challenges in Commodity Markets and on Raw Materials" (opinion paper, CCMI 091, European Economic and Social Committee, Brussels, accessed December 3, 2014), http://www.eesc.europa.eu/?i=portal.en.ccmi-opinions.15877. This list included antimony, gallium, platinum group metals, beryllium, germanium, phosphate rock, borates, indium, heavy REEs, chromium,

magnesite, light REEs, cobalt, magnesium, silicon metal, coking coal, natural graphite, tungsten, fluorspar, niobium. According to the EU report, antimony, fluorspar, gallium, germanium, graphite, indium, magnesium, rare earths, tungsten all came from China.

6. Estimates vary in the literature upward to 97%.

7. Cindy Hurst, "China's Rare Earth Elements Industry: What Can the West Learn?" (report Institute for the Analysis of Global Security, "IAGS," Washington DC, March 2010), accessed August 20, 2011, http://fmso.leavenworth.army.mil/documents/rareearth.pdf.

8. Justin McCurry, "Japan-China Row Escalates over Fishing Boat Collision," *The Guardian*, September 9, 2010, accessed June 3, 2011, https://www.theguardian.com/world/2010/sep/09/japan-china-fishing-boat-collision; Keith Bradsher, "Amid Tension, China Blocks Vital Exports to Japan," *New York Times*, September 22, 2010, accessed November 2, 2013, http://www.nytimes.com/2010/09/23/business/global/23rare.html?pagewanted=all&_r=1; William Wan, "Boat Collision Sparks Anger, Breakdown in China-Japan Talks," *Washington Post*, September 20, 2010, accessed January 25, 2013, http://www.washingtonpost.com/wp-dyn/content/article/2010/09/20/AR2010092000130.html.

9. For a detailed account and interpretations of the incident, both within a dominant narrative of "China's rise" and "Japan's decline," together with an opposing reconstruction in order to question the epistemological/constructivist nature of international relations (IR) narratives, see Linus Hagström, "'Power Shift' in East Asia? A Critical Reappraisal of Narratives on the Diaoyu/Senkaku Islands Incident in 2010," *Chinese Journal of International Politics* 5, no. 3 (2012): 267–97, doi:10.1093/cjip/pos011.

10. Ian Johnson, "China and Japan Bristle over Disputed Chain of Islands," *New York Times*, September 8, 2010, accessed January 10, 2012, http://www.nytimes.com/2010/09/09/world/asia/09beijing.html; Mark Landler, "U.S. Works to Ease China-Japan Conflict," *New York Times*, October 30, 2010, accessed January 10, 2012, http://www.nytimes.com/2010/10/31/world/asia/31diplo.html.

11. The following *New York Times* article was widely reproduced in Western media, fueling geopolitical anxieties with the added worry that the embargo would be extended beyond Japan. Keith Bradsher cites "anonymous rare earth officials" to claim that Chinese customs officials were holding up shipments to the US as well; Keith Bradsher, "China Said to Expand Rare Earths Embargo to West," *New York Times*, October 19, 2010, http://www.nytimes.com/2010/10/20/business/global/20rare.html.

12. Data on rare earths, published in January 2012 by the US Geological Survey, Mineral Commodity Summaries, shows the steep increase in refined rare-earth imports into the United States between 2010 and 2011. It also provides a breakdown of the distribution of rare earths by end use. "In 2011, rare earths were not mined in the United States; . . . the United States continued to be a major consumer, exporter, and importer of rare-earth products in 2011. The estimated value of refined rare earths imported by the United States in 2011 was $696 million, an increase from $161 million imported in 2010. Based on reported data through

August 2011, the estimated 2011 distribution of rare earths by end use, in decreasing order, was as follows: catalysts, 47%; metallurgical applications and alloys, 13%; alloys, 11%; glass polishing and ceramics, 10%; permanent magnets, 9%; ceramics, 5%; rare-earth phosphors for computer monitors, lighting, radar, televisions, and x-ray-intensifying film, 5%." See "Rare Earths," USGS, Mineral Commodity Summary, Washington, DC, January 2012, accessed June 2, 2012, http://minerals.usgs.gov/minerals/pubs/commodity/rare_earths/mcs-2012-raree.pdf.

13. A few scholars and analysts have expressed doubts about whether such an embargo ever took place, providing some limited data on port shipments during that period. The particular data itself (see, for example, Alastair Iain Johnston's analysis, cited in chapter 1, section Economic Statecraft) does not support the claim, but in addition lacks sufficient detail to evaluate what kind of rare earths were in fact curtailed, since it does not separate more valuable heavy rare earths from light rare earths, which are more abundant and less critical. Regardless of these discrepancies, what must be taken under consideration, not just in this particular instance but also more widely in international relations, is the perception of the threat and the vocal response of world leaders to what was characterized as a provocation on the part of the Chinese. The high-level meetings, as well as Hillary Clinton's Asia tour to address the problem, would not have occurred except as part of a pushback on China's muscle flexing in the region. Alastair Iain Johnston, "How New and Assertive Is China's New Assertiveness?," *International Security* 37, no. 4 (2013): 7–48.

14. "Joint Press Availability with Japanese Foreign Minister Seiji Maehara," US Department of State, October 27, 2010, accessed January 21, 2013, https://2009-2017.state.gov/secretary/20092013clinton/rm/2010/10/150110.htm.

15. "Joint Press Availability with Japanese Foreign Minister Seiji Maehara."

16. "China Cuts Rare Earth Export Quotas 40%, Newspaper Says," *Mineweb*, August 11, 2010, accessed January 2, 2012, http://www.mineweb.com/archive/china-cuts-rare-earth-export-quotas-40-newspaper-says/.

17. In chapter 4, "The World Reacts to the Realization of the Growing Scarcity of REEs," I argue that such diversification has yet to happen and that it is not clear whether either the political will or the coordination among China's competitors is sufficient for it to be accomplished.

18. Eugene Gholz, "Rare Earth Elements and National Security" (report, Council on Foreign Relations, Washington, DC, October 2014), accessed January 4, 2014, i.cfr.org/content/publications/attachments/Energy%20Report_Gholz.pdf.

19. Johnston, "How New and Assertive?," 7–48.

20. Tom Miles, "China Loses Appeal of WTO Ruling on Exports of Rare Earths," *Reuters*, August 7, 2014, accessed September 1, 2014, http://www.reuters.com/article/us-china-wto-rareearths-idUSKBN0G71QD20140807; "WTO | Dispute Settlement—the Disputes—DS431," May 2015, accessed September 1, 2015, https://www.wto.org/english/tratop_e/dispu_e/cases_e/ds431_e.htm.

21. "Molycorp Stock Price," *Bloomberg.com*, http://www.bloomberg.com/quote/MCP:US.

22. "Lynas Stock Price," *Bloomberg.com*, http://www.bloomberg.com/quote/LYC:AU.

23. John W. Miller and Anjie Zheng, "Molycorp Files for Bankruptcy Protection," *Wall Street Journal*, June 25, 2015, accessed July 22, 2015, http://www.wsj.com/articles/SB10907564710791284872504581069270334872848.

24. Dingding Chen, Xiaoyu Pu, and Alastair Iain Johnston, "Correspondence: Debating China's Assertiveness," *International Security* 38, no. 3 (Winter 2013/14): 176–83.

25. "Green Growth Strategy for Energy: A Window of Opportunity" (report by the OECD and International Energy Agency [IEA], Paris, December 2011), accessed February 2, 2013, http://www.oecd.org/greengrowth/greening-energy/49157149.pdf; "2017 Outlook for Energy: A View to 2040" (report, ExxonMobil, accessed June 1, 2017, http://corporate.exxonmobil.com/en/energy/energy-outlook.

26. Rawi Abdelal and Adam Segal, "Has Globalization Passed Its Peak?" in *International Politics, Enduring Concepts and Contemporary Issues*, ed. Robert J. Art and Robert Jervis (New York: Pearson/Longman, 2009), 344–45.

27. An example that highlights this type of strategy was OPEC's decision between 2014 and 2016 not to decrease output but to let oil prices plummet. Though painful to many of its members, this strategy was not only designed to put pressure on the US fracking industry, but also to make new exploration unappealing at such low prices. In essence, OPEC was attempting to maintain its market share by riding the wave of low prices and discouraging new product to come online. This strategy may have reflected a future concern that with climate change worsening, existing resources of fossil fuels arguably will eventually need to stay underground to control the temperature rise.

28. Hurst, "China's Rare Earth Elements Industry."

29. "Rare Earths: Strategic Inputs to Sustainability," Institute français des relations internationals (IFRI), May 20, 2010, accessed February 14, 2012, https://www.ifri.org/en/debats/rare-earths-strategic-inputs-sustainability.

## CHAPTER 1

1. Vinod K. Aggarwal and Sara A. Newland, eds., *Responding to China's Rise*, Political Economy of the Asia Pacific (Cham, Switzerland: Springer International, 2015), 1–12.

2. Philippe Le Billon, "The Geopolitical Economy of Resource Wars," in *The Geopolitics of Resource Wars*, ed. Philippe Le Billon (London: Frank Cass, 2005), 3.

3. Le Billon, "Geopolitical Economy of Resource Wars," 2. Cf. Michael Klare, *Resource Wars: The New Landscape of Global Conflict* (New York: Metropolitan, 2002).

4. Ronnie Lipschutz, *When Nations Clash: Raw Materials, Ideology, and Foreign Policy* (New York: Ballinger, 1989), 1–7; Thomas Homer-Dixon, *Environment, Scarcity, and Violence* (Princeton, NJ: Princeton University Press, 1999), 137.

5. Abdelal and Segal, "Has Globalization Passed Its Peak?," in Art and Jervis, *International Politics*, 344–45. Abdelal and Segal write:

    Oil has become the global commodity, unparalleled in importance . . . Throughout Latin America, governments have reasserted their authority over extraction projects that they once had ceded to foreign firms . . . In response, a number of oil have-nots have taken measures to insulate themselves from a disruption in their oil supply. This helps explain China's seemingly illogical drive to acquire stakes in oil production facilities abroad. So long as oil remains a global commodity, consumers need not own the means of its production: they can simply buy all they need on the world market. China, however, seems to be preparing for a day when oil becomes far harder to acquire and transport and has thus signed various oil and natural-gas agreements . . . with Angola, Brazil, Iran, Nigeria, Venezuela, and Sudan. This strategy makes so little economic sense that it can only be explained by an expectation that global oil markets will at some point break down, due to either a worldwide recession or conflict between China and the United States.

6.   John E. Tilton, *On Borrowed Time? Assessing the Threat of Mineral Depletion* (Washington, DC: RFF Press, 2003), 1.

7.   Aaron Greenfield and T. E. Graedel, "The Omnivorous Diet of Modern Technology," *Resources, Conservation and Recycling* 74 (May 2013): 1–7.

8.   Greenfield and Graedel, "Omnivorous Diet," 1–7.

9.   Wensheng Cao and Christoph Bluth, "Challenges and Countermeasures of China's Energy Security," *Energy Policy* 53 (February 2013): 381–88.

10.  Tilton, *On Borrowed Time?*, 1.

11.  Dianzuo Wang, "Perspectives on China's Mining and Mineral Industry," in *A Review on Indicators of Sustainability for the Mineral Extraction Industries*, ed. Roberto C. Villas-Boas, Deborah Shields, et al. (Rio de Janeiro: CYTED, 2005), 105–13.

12.  Ecorys, "Mapping Resource Prices: The Past and the Future" (report for the European Commission, Ecorys, Rotterdam, Netherlands, October 25, 2012) accessed September 20, 2014, http://ec.europa.eu/environment/enveco/resource_efficiency/pdf/summary_mapping_resource_prices.pdf.

13.  Ecorys, "Mapping Resource Prices."

14.  Raghuram Rajan, "The Great Game Again?," *Finance and Development* 43, no. 4 (2006): 54–55.

15.  "When we refer to deposits we are speaking of concentrations of minerals that have a strong economic interest. Deposits contain ores. What is not ore is known as common rock. Ores are classified into resources and reserves. Resources are defined as concentrations of material from which it is feasible to economically extract a mineral commodity. Reserves are a subset of identified resources that can be profitably exploited given current prices and the technology that exists. This category can shift if the price of a mineral drops precipitously making it economically uninteresting to extract. Reserves then are reclassified as resources." Deborah J. Shields and Slavko V Šolar, "Responses to Alternative Forms of Mineral Scarcity: Conflict and Cooperation," in *Beyond Resource Wars: Scarcity, Environmental Degradation, and International Cooperation*, ed. Shlomi Dinar (Cambridge, MA: MIT Press, 2011), 246.

16.  Tilton, *On Borrowed Time?*, 612–13.

17.  In the beginning of the twentieth century, circa 1910, there was also a fear of mineral resource exhaustion. John Tilton underscores this by quoting Gifford Pinchot, a leader of the American Conservation Movement, who expressed these concerns as follows: "our supplies of iron ore, mineral oil, and many of the great fields are already exhausted. Mineral resources such as these when once gone are gone forever." John E. Tilton, "Exhaustible Resources and Sustainable Development: Two Different Paradigms," *Resource Policy* 22, no. 1/2 (1996): 91–97.

18.  Thomas Malthus, *An Essay on the Principle of Population* (London: J. Johnson, 1798).

19.  Tilton, "Exhaustible Resources and Sustainable Development," 91–97.

20.  J. E. Young, "Mining the Earth," in *State of the World*, ed. Lester R. Brown, Edward C. Wolf, and Linda Stark (New York: W. W. Norton, 1992), 110–18.

21.  Tilton, "Exhaustible Resources and Sustainable Development," 91–97.

22.  Stephen E. Kesler, *Mineral Resources, Economics and the Environment* (New York: Macmillan, 1996).

23.  Julian L. Simon, *The State of Humanity* (Cambridge, MA: Blackwell, 1995), 10–11.

24.  Tilton, "Exhaustible Resources and Sustainable Development," 91–97.

25.  Harold J. Barnett and Chandler Morse, *Scarcity and Growth* (Baltimore: Johns Hopkins Press for Resources for the Future, 1963).

26.  Shields and Šolar, "Responses to Alternative Forms of Mineral Scarcity," in Dinar, *Beyond Resource Wars*, 251.

27.  Cf. Ester Boserup, *The Conditions of Agricultural Growth: The Economics of Agrarian Change under Population Pressure* (London: Allen & Unwin, 1965); Julian L. Simon, *The Ultimate Resource 2* (Princeton, NJ: Princeton University Press, 1996); Bjørn Lomborg, "Resource Constraints or Abundance?" in *Environmental Conflict*, ed. Paul F. Diehl and Nils Petter Gleditsch (Boulder, CO: Westview, 2001), 125–54; Kristine Juul, "Transhumance, Tubes, and Telephones: Drought-Related Migration as a Process of Innovation," in *Beyond Territory and Scarcity: Exploring Conflicts over Natural Resource Scarcities*, ed. Quentin Gausset, Michael Whyte, and Michael Birch-Thomsen (Uppsala, Sweden: Nordic Africa Institute, 2005), 112–34; Michael Mortimore, "Social Resilience in African Dryland Livelihoods: Deriving Lessons for Policy," in Gausset, Whyte, and Birch-Thomsen, *Beyond Territory and Scarcity*, 46–69.

28.  Thomas Homer-Dixon, "Cornucopians and Neo-Malthusians," in Art and Jervis, *International Politics*, 522.

29.  Homer-Dixon, "Cornucopians and Neo-Malthusians," in Art and Jervis, *International Politics*, 522.

30.  Homer-Dixon, "Cornucopians and Neo-Malthusians," in Art and Jervis, *International Politics*, 523–24.

31.  R. B. Gordon, M. Bertram, and T. E. Graedel, "On the Sustainability of Metal Supplies: A Response to Tilton and Lagos," *Resources Policy* 32, no. 1/2 (2007): 24–28.

32.  Gordon, Bertram, and Graedel, "On the Sustainability of Metal Supplies," 24–28.

33.  Gordon, Bertram, and Graedel, "On the Sustainability of Metal Supplies," 24–28.

34.  Thomas Princen, *The Logic of Sufficiency* (Cambridge, MA: MIT Press, 2005).

35.  Markus A. Reuter and Antoinette van Schaik, "Transforming the Recovery and Recycling of Nonrenewable Resources," in *Linkages of Sustainability*, ed. Thomas E. Graedel and Ester van der Voet (Cambridge, MA: MIT Press, 2010), 150–62.

36.  National Resources Council, *Minerals, Critical Minerals, and the U.S. Economy* (report, Washington, DC: National Academies Press, 2008).

37.  Shields and Šolar, "Responses to Alternative Forms of Mineral Scarcity," in Dinar, *Beyond Resource Wars*, 253–55.

38.  Robert Mandel, *Conflict over the World's Resources: Background, Trends, Case Studies, and Considerations for the Future* (New York: Greenwood, 1988), 79. For

instance, the Soviets imposed a crippling embargo on strategic minerals to China in the 1960s as a result of the Sino-Soviet rift.

39.   Mandel, *Conflict over the World's Resources.*

40.   "The fundamental source of the resource conflict between the United States and the Soviet Union over strategic minerals is geopolitics, in the sense that a struggle exists over security of access to geographically concentrated raw materials critical to international power and influence." Mandel, *Conflict over the World's Resources,* 79. According to Haglund, "It is only in the 20th century that minerals have appeared as a reason for, not merely a means of, fighting." David G. Haglund, "The New Geopolitics of Minerals: An Inquiry into the Changing International Significance of Strategic Minerals," *Political Geography Quarterly* 5 (1986): 221–40.

41.   Embargoes have been used repeatedly and for different reasons. Recent examples include the embargo on Iran in response to the latter's nuclear program and the 1973–74 oil embargo in response to the Arab-Israeli war. In a speech at the Brookings Institute, in November 2012, Hilary Clinton spoke about the economic impact of the most recent sanctions on Iran. "Iran's oil production is down a million barrels a day. That costs the Iranian government $3 billion every month." Hillary Clinton, "U.S. and Europe: A Revitalized Global Partnership" (speech, Brookings Institute, Washington, DC, November 29, 2013) accessed February 14, 2013, http://www.brookings.edu/~/media/events/2012/11/29%20clinton/20121129_transatlantic_clinton.pdf.

42.   Lesley Wroughton and Yeganeh Torbati, "Nuclear Sanctions Lifted as Iran, U.S. Agree on Prisoner Swap," *Reuters,* January 17, 2016, accessed February 12, 2016, http://www.reuters.com/article/us-iran-nuclear-zarif-idUSKCN0UU0C7. In January 2016, world powers lifted crippling sanctions against the Islamic Republic in return for Tehran complying with a deal to curb its nuclear ambitions. The United States continued to maintain less comprehensive sanctions over Iran's missile program. It did, however, lift banking, steel, shipping, and other sanctions.

43.   "Iran Sanctions," Bureau of Public Affairs, Department of State, Office of Website Management, November 1, 2012, accessed December 5, 2012, http://www.state.gov/e/eb/tfs/spi/iran/index.htm; Office of Foreign Assets Control (OFAC), "Iranian Transactions and Sanctions Regulations," (advisory memorandum, US Department of Treasury, Washington, DC, January 10, 2013), accessed February 4, 2013, http://www.treasury.gov/resource-center/sanctions/Programs/Documents/20130110_iran_advisory_exchange_house.pdf.

44.   OFAC, "Iranian Transactions and Sanctions Regulations."

45.   Yergin, writing back in 2008, claimed, "The control of Venezuela's oil wealth permitted Chavez to expand his influence over Latin America and pursue his agenda of 'socialism for the 21st century' across the world stage." Daniel Yergin, *The Prize* (New York: Free Press, 2008), 769.

46.   Reid W. Click and Robert J. Weiner, "Resource Nationalism Meets the Market: Political Risk and the Value of Petroleum Reserves," *Journal of International Business Studies* 41, no. 5 (2010): 783–803.

47.   "The core idea behind a resource nationalism perspective is that the natural resources in the ground and under the sea are a 'national patrimony' and

consequently should be used for the benefit of the nation rather than for private gain. In addition, the commodity itself has an intrinsic value, not one determined by the market, and this value belongs to the nation." David R. Mares, "Resource Nationalism and Energy Security in Latin America: Implications for Global Oil Supplies" (working paper, James A. Baker III Institute for Public Policy, Houston, TX, January 2010), accessed June 12, 2010, http://www.bakerinstitute.org/media/files/Research/edacf0ea/EF-pub-MaresResourceNationalismWorkPaper-012010.pdf.

48.  Homer-Dixon, *Environment, Scarcity and Violence*, 12.

49.  Richard M. Auty, "The Political Economy of Resource-Driven Growth," *European Economic Review* 45, no. 4–6 (2001): 839–46.

50.  Jeffrey D. Sachs and Andrew M. Warner, "The Curse of Natural Resources," *European Economic Review* 45, no. 4–6 (2001): 827–38.

51.  Cf. Colin Kahl, *States, Scarcity, and Civil Strife in the Developing World* (Princeton, NJ: Princeton University Press, 2008).

52.  Donald Greenlees, "Indonesian Court Acquits Newmont Mining," *New York Times*, April 25, 2007, accessed February 12, 2013, http://www.nytimes.com/2007/04/25/world/asia/25indo.html.

53.  Theodore E. Downing, *Avoiding New Poverty: Mining-Induced Displacement and Resettlement*, vol. 52 (report, International Institute for Environment and Development, London, UK, 2002).

54.  Homer-Dixon, *Environment, Scarcity and Violence*, 144.

55.  Paul Collier and Anke Hoeffler, "Greed and Grievance in Civil War," *Oxford Economic Papers* 56, no. 4 (2004): 563–95.

56.  David A. Baldwin, *Economic Statecraft* (Princeton, NJ: Princeton University Press, 1985), 20.

57.  Baldwin, *Economic Statecraft*, 41.

58.  Daniel W. Drezner, *The Sanctions Paradox: Economic Statecraft and International Relations* (Cambridge: Cambridge University Press, 1999), 7.

59.  The contested islands are particularly coveted by both sides because of their richness in important resources that include minerals as well as oil.

60.  Michael Wines, "China's Leader Says He Is 'Worried' over U.S. Treasuries," *New York Times*, March 13, 2009, accessed September 20, 2016, http://www.nytimes.com/2009/03/14/business/worldbusiness/14china.html.

61.  Zhou Xiaochuan, "Reform the International Monetary System," *BIS Review* 41 (2009): 1–3; Malcolm Moore, "Top Chinese Banker Guo Shuqing Calls for Wider Use of Yuan," *Telegraph* (London), June 8, 2009, accessed September 10, 2016, http://www.telegraph.co.uk/finance/financialcrisis/5473491/Top-Chinese-banker-Guo-Shuqing-calls-for-wider-use-of-yuan.html.

62.  Cf. Nina Hachigian, ed., *Debating China: The U.S.-China Relationship in Ten Conversations* (Oxford: Oxford University Press, 2014); Chi Wang, *Obama's Challenge to China: The Pivot to Asia* (Farnham, UK: Ashgate, 2015); Thomas J. Christensen, *The China Challenge* (New York: W. W. Norton, 2015), 3.

63.  "EU, US Preparing WTO Action against China: Source," *Terra Daily*, June 11, 2009, accessed May 20, 2012, http://www.terradaily.com/reports/EU_US_preparing_

WTO_action_against_China_source_999.html; Jill Jusko, "Raw Material Risks," *Industry Week*, October 21, 2009, accessed September 26, 2016, http://www.industryweek.com/articles/raw_material_risks_20203.aspx.

64. Christensen, *China Challenge*, 246.
65. Johnston, "How New and Assertive?," 7–48.
66. Christensen, *China Challenge*, 261.
67. Michèle Flournoy and Janine Davidson, "Obama's New Global Posture: The Logic of U.S. Foreign Deployments," *Foreign Affairs* 91, no. 4 (2012): 54–63.
68. Joseph. S. Nye Jr., *Soft Power: The Means to Success in World Politics* (New York: Public Affairs, 2004), 4.
69. In the State of the Union Address (2013), President Obama declared his commitment to conclude the Trans-Pacific Partnership in order to achieve economic integration between the US and Asian countries, "to boost American exports, support American jobs, and level the playing field in the growing markets of Asia, we intend to complete negotiations on a Trans-Pacific Partnership." See "Transcript of Obama's State of the Union Address," *ABC News*, February 13, 2013, accessed February 14, 2013, http://abcnews.go.com/Politics/OTUS/transcript-president-barack-obamas-2013-state-union-address/story?id=18480069; The initial list of countries that had agreed to participate included: Singapore, Chile, New Zealand, Brunei, Australia, Peru and Vietnam, but others had also expressed interest at the time. The US government stated that, "The initiative starts with an economically-significant group of countries that share our vision of negotiating a high-standard, 21st century regional agreement. The goal is to include additional Asia-Pacific countries in successive clusters to eventually cover a region that represents more than half of global output and over 40 percent of world trade." "The Trans-Pacific Partnership," Office of the United States Trade Representative, USTR.gov, accessed September 26, 2016, https://ustr.gov/sites/default/files/TPPFAQ.pdf.

The Obama administration's focused push to further expand the TPP had been carefully monitored by China. A number of Chinese scholars at the time expressed their concerns on the subject. Xiangyang Li, director of the Institute of Asia-Pacific Studies under the Chinese Academy of Social Sciences (CASS), for instance, had argued that "the TPP is an important component of the U.S. strategy of 'Returning to Asia' that includes both economic and geopolitical incentives, and one of its major incentives is to contain China's rise." Accordingly, Li had forewarned that when the TPP came into force, it would "seriously undermine the effectiveness of the APEC framework, and China's being excluded from the TPP will undercut the East Asian regional integration process that China has been propelling for over a decade, posing a great challenge to China's rise in the future." Li was not alone in his analysis of the possible impacts of the TPP on China. Yang Jiemian, president of the Shanghai Institute of International Studies, wrote that the "US 'dilutes' and 'reduces' (rather than "contains") China's influence in the Asia-Pacific region, which could be seen as a 'soft confrontation.'" Guoyou Song and Wen Jin Yuan, "China's Free Trade Agreement Strategies," *Washington Quarterly* 35, no. 4 (2012): 107–19; Evelyn S. Devadason, "The Trans-Pacific Partnership (TPP): The Chinese Perspective," *Journal of Contemporary China* 23, no. 87 (2014): 462–79;

Yang Jiemian, "The Change of America's Power and Re-structure of International System," *China International Studies* (March/April 2012), http://www.cssn.cn/upload/2013/02/d20130228095946338.pdf; Weng Jin Yuan, "The Trans-Pacific Partnership and China's Corresponding Strategies" (briefing report, Center for Strategic and International Studies, Washington, DC, June 2012), accessed on February 15, 2013, https://csis-prod.s3.amazonaws.com/s3fs-public/legacy_files/files/publication/120620_Freeman_Brief.pdf.

Later in the chapter, I note that after Donald Trump's election, the United States withdrew from the TPP in line with electoral promises by the new administration. This has not diminished the US pre-occupation with China and Asia overall.

70. Christensen, *China Challenge*, 250.

71. Robert M. Hathaway and Wilson Lee, eds., "George W. Bush and East Asia: A First Term Assessment" (report, Woodrow Wilson International Center for Scholars, Washington, DC, 2005), accessed August 3, 2013, https://www.wilsoncenter.org/sites/default/files/bushasia2rpt.pdf.

72. "U.S.-China Strategic and Economic Dialogue," Bureau of Public Affairs Department of State, July 29, 2009, accessed August 30, 2016, http://www.state.gov/e/eb/tpp/bta/sed/.

73. "Remarks by President Obama to the Australian Parliament," Whitehouse.gov, November 17, 2011, accessed August 1, 2016, https://www.whitehouse.gov/the-press-office/2011/11/17/remarks-president-obama-australian-parliament.

74. Flournoy and Davidson, "Obama's New Global Posture," 54–63.

75. Hillary Clinton, "America's Pacific Century" (remarks, US Department of State, Washington, DC, November 10, 2011) accessed on March 1, 2016, http://www.state.gov/secretary/20092013clinton/rm/2011/11/176999.htm.

76. There is an extensive literature available on China's rise and its significance for the United States. Richard K. Betts and Thomas J. Christensen, "China: Getting the Questions Right," *National Interest*, no. 62 (2000): 17–29; Aaron L. Friedberg, *A Contest for Supremacy: China, America, and the Struggle for Mastery in Asia* (New York: W. W. Norton, 2012); Rosemary Foot, *China, the United States, and Global Order* (New York: Cambridge University Press, 2011).

77. Cf. Stephen S. Roach, *Unbalanced: The Codependency of America and China* (New Haven, CT: Yale University Press, 2014). This overt shift of American focus toward Asia has EU policymakers worrying about the future of transatlantic cooperation. Europe's engagement with China has been primarily economic, but its relation to the United States has been political, economic, and security oriented. Javier Solana, former secretary-general of NATO and the EU's high representative for the Common Foreign and Security Policy, wrote that a transatlantic free trade agreement would be crucial for the EU. Specifically, he underscored that "if current trends continue, Asia could soon surpass North America and Europe in global power. It will have a higher GDP, larger population, higher military spending, and more technological investment. In this geopolitical context, Europe and the US need each other more than ever, making greater transatlantic co-operation crucial." Furthermore, according to Solana, in a recent speech at the Brookings Institute, Secretary Clinton extended an invitation to the EU to join the US in

its reorientation toward Asia so that "Asia is seen not only as a market, but also as a focus of common strategic action." European preoccupation with China has thus far mainly revolved around economic opportunity, trade, and competition. See Javier Solana, "Transatlantic Free Trade?," *Politico*, January 4, 2013, accessed February 14, 2013 http://www.politico.eu/article/transatlantic-free-trade/; Sophia Kalantzakos, *EU, US and China Tackling Climate Change: Policies and Alliances for the Anthropocene* (Abingdon, UK: Routledge, 2017), 1–15.

78.  "Sustaining U.S. Global Leadership: Priorities for 21st Century Defense" (strategic guidance report, U.S. Department of Defense, Washington, DC, January 5, 2012), accessed May 10, 2013, http://www.defense.gov/news/Defense_Strategic_Guidance.pdf.

79.  "Sustaining U.S. Global Leadership."

80.  "Sustaining U.S. Global Leadership."

81.  Brahma Chellaney, "US Strategy in the Asia-Pacific" (report, Aljazeera Center for Studies, Doha, Qatar, February 2012), accessed September 16, 2012, http://studies.aljazeera.net/mritems/Documents/2012/2/15/201221510538465734U.S.%20Strategy%20in%20the%20Asia-Pacific.pdf.

82.  Christensen, *China Challenge*, xiv–xv.

83.  Paul Kennedy, *The Rise and Fall of the Great Powers: Economic Change and Military Conflict from 1500 to 2000* (New York: Vintage, 1987), 447–58. As China rises there is a growing need to consider a non-Western view on international politics. Today's IR theories are widely regarded as a product of American social science. Since the 1980s, there has been an ongoing and robust academic discussion within China to develop its own Chinese School of IR. See Junxian Gan, "The Analysis of 'China's Responsibility' and Its Diplomatic Countermeasures," *Global Review*, April 2014, accessed December 7, 2016, http://en.cnki.com.cn/Article_en/CJFDTOTAL-GJZW201004016.htm. "China's Responsibility and 'China Responsibility Theory,'" *International Studies*, March 2007, accessed December 7, 2016, http://en.cnki.com.cn/Article_en/CJFDTOTAL-GJWY200703000.htm; Hachigian, ed., *Debating China*.

84.  Tom Phillips, "Beijing Summons US Ambassador over Warship in South China Sea," *The Guardian*, October 27, 2015, accessed February 10, 2016, https://www.theguardian.com/world/2015/oct/27/us-warship-lassen-defies-beijing-sail-disputed-south-china-sea-islands.

85.  "12 Pacific Nations Sign Major Free Trade Agreement," *Voice of America* (VOA), February 4, 2016, accessed February 10, 2016, http://www.voanews.com/a/twelve-pacific-nationsfree-trade-agreement-tpp/3176187.html.

86.  Peter Baker, "Trump Abandons Trans-Pacific Partnership, Obama's Signature Trade Deal," *The New York Times*, January 23, 2017, sec. Politics, https://www.nytimes.com/2017/01/23/us/politics/tpp-trump-trade-nafta.html.

87.  Ankit Panda, "Straight From the US State Department: The 'Pivot' to Asia Is Over," *The Diplomat*, accessed June 24, 2017, http://thediplomat.com/2017/03/straight-from-the-us-state-department-the-pivot-to-asia-is-over/.

88.  "U.S. Warship Drill Meant to Defy China's Claim over Artificial Island: Officials," *Reuters*, May 26, 2017, http://www.reuters.com/article/us-usa-southchinasea-navy-idUSKBN18K353.

89. Chien-Peng Chung, "Japan's Involvement in Asia-Centered Regional Forums in the Context of Relations with China and the United States," *Asian Survey* 52, no. 3 (2011): 407–28.

90. Yohei Kono, "Japan's Role in Asia-Pacific Regional Cooperation" (speech, Japan National Press Club, July 28, 1995) accessed September 26, 2016, http://www.mofa.go.jp/announce/announce/archive_3/asia.html.

91. Attributed to former Japanese prime minister Hashimoto Ryutaro and launched in 1997, Japan named this "Silk Road diplomacy." It was meant to signify a closer collaboration with Central Asia. Japan's interest in the wider region was attributed to its desire to both diversify its supplies of oil and gas, given how resource poor the country is, and for geostrategic considerations since the area is viewed as Japan's backyard. The policy of investing in the region was coupled with high levels of development assistance.

92. Muthiah Alagappa, "Japan's Political and Security Role in the Asia-Pacific Region," *Contemporary Southeast Asia* 10, no. 1 (1998): 17–54.

93. Christopher B. Johnstone, "Paradigms Lost: Japan's Asia Policy in a Time of Growing Chinese Power," *Contemporary Southeast Asia* 21, no. 3 (1999): 365–85.

94. Yul Sohn, "Japan's New Regionalism: China Shock, Values, and the East Asian Community," *Asian Survey* 50, no. 3 (2010): 497–519.

95. Narushige Michishita and Richard J. Samuels, "Hugging and Hedging: Japanese Grand Strategy in the Twenty-First Century," in *Worldviews of Aspiring Powers: Domestic Foreign Policy Debates in China, India, Iran, Japan and Russia*, ed. Henry R. Nau and Deepa Ollapally (New York: Oxford University Press, 2012), 147–73.

96. Martin Fackler and Ian Johnson, "Sleepy Islands and a Smoldering Dispute," *New York Times*, September 20, 2012, accessed December 9, 2012, http://www.nytimes.com/2012/09/21/world/asia/japan-china-trade-ties-complicate-island-dispute.html.

97. Martin Fackler, "Japan Expands Its Regional Military Role," *New York Times*, November 26, 2012, accessed November 27, 2012, http://www.nytimes.com/2012/11/27/world/asia/japan-expands-its-regional-military-role.html.

98. "The Constitution of Japan," accessed August 1, 2016, http://www.japanese-lawtranslation.go.jp/law/detail_main?id=174; Sayuri Umeda, "Japan: Article 9 of the Constitution | Law Library of Congress," February 2006, accessed August 1, 2016, https://www.loc.gov/law/help/japan-constitution/article9.php.

99. Matt Ford, "Japan Curtails Its Pacifist Stance," *The Atlantic*, September 19, 2015, accessed August 1, 2016, http://www.theatlantic.com/international/archive/2015/09/japan-pacifism-article-nine/406318/.

100. "Japan Passes Changes to Pacifist Constitution to Allow Troops to Fight Abroad," *ABC News*, September 19, 2015, accessed December 1, 2015, http://www.abc.net.au/news/2015-09-19/japan-parliament-passes-change-to-pacifist-constitution/6788456.

101. Sheila Smith, "Will Japanese Change Their Constitution?," *Forbes*, July 7, 2016, accessed August 1, 2016, http://www.forbes.com/sites/sheilaasmith/2016/07/07/japans-constitution/.

102. Smith, "Will Japanese Change Their Constitution?"
103. Commission of the European Communities, "A Long-Term Policy for China-Europe Relations" (COM (1995) 279/final, Communication of the Commission, Brussels, July 5, 1995), accessed January 2015, http://eeas.europa.eu/china/docs/com95_279_en.pdf.
104. Commission of the European Communities, "Long-Term Policy for China-Europe Relations."
105. "China—Trade—European Commission," European Commission, accessed September 26, 2016, http://ec.europa.eu/trade/policy/countries-and-regions/countries/china/.
106. Nicola Casarini, *Remaking Global Order: The Evolution of Europe-China Relations and Its Implications for East Asia and the United States* (New York: Oxford University Press, 2009).
107. Neill Nugent, *The Government and Politics of the European Union* (Basingstoke, UK: Palgrave Macmillan, 2010), 3–17.
108. "The EU in the World: The Foreign Policy of the European Union," June 2007, accessed January 20, 2013, http://ec.europa.eu/publications/booklets/move/67/en.pdf.
109. Asteris Huliaras, "The Illusion of Unitary Players and The Fallacy of Geopolitical Rivalry: The European Union and China in Africa," *Round Table: The Commonwealth Journal of International Affairs* 101, no. 5 (2012): 425–34; Steven Hill, *Europe's Promise: Why the European Way Is the Best Hope in an Insecure Age* (Berkeley: University of California Press, 2010).
110. "Green Paper on the Security of Energy Supply," European Commission, November 29, 2000, accessed February 10, 2011, http://eur-lex.europa.eu/legal-content/EN/TXT/HTML/?uri=URISERV:l27037&from=EN; "EU in the World: The Foreign Policy of the European Union."
111. "EU in the World: The Foreign Policy of the European Union."
112. Joshua Eisenman, Eric Heginbotham, and Derek Mitchell, *China and the Developing World: Beijing's Strategy for the Twenty-First Century* (Armonk: M. E. Sharpe, 2007), ix. Cf. Jonathan Spence, *The Gate of Heavenly Peace: The Chinese and Their Revolution, 1895–1980* (New York: Penguin, 1981).
113. Evan S. Medeiros and M. Taylor Fravel, "China's New Diplomacy," *Foreign Affairs* 82, no. 6 (2003): 22–35.
114. Winberg Chai, "The Ideological Paradigm Shifts of China's World Views: From Marxism-Leninism-Maoism to the Pragmatism-Multilateralism of the Deng-Jiang-Hu Era," *Asian Affairs* 30, no. 3 (Fall 2003): 163–75.
115. Eisenman, Heginbotham, and Mitchell, *China and the Developing World*, x.
116. Andrew Kohut, "How the World Sees China," Pew Research Center's Global Attitudes Project, December 11, 2007, accessed May 10, 2012, http://www.pewglobal.org/2007/12/11/how-the-world-sees-china/.
117. Kohut, "How the World Sees China."
118. "Chapter 2: China's Image," Pew Research Center's Global Attitudes Project, July 14, 2014, accessed September 16, 2016, http://www.pewglobal.org/2014/07/14/chapter-2-chinas-image/.

119. Roderick MacFarquhar, *The Politics of China: Sixty Years of the People's Republic of China* (Cambridge: Cambridge University Press, 2011).

120. Ming Xia, "'China Threat' or a 'Peaceful Rise of China'?," *New York Times*, March 24, 2006, accessed July 25, 2012, http://www.nytimes.com/ref/college/coll-china-politics-007.html.

121. Eisenman, Heginbotham, and Mitchell, *China and the Developing World*, xiv.

122. Christian Constantin, "Understanding China's Energy Security," *World Political Science Review* 3 (2007): 1–30.

123. Sigfrido Burgos Cáceres, and Sophal Ear, "Geopolitics of China's Global Resources Quest," *Geopolitics* 17, no. 1 (2012): 47–79.

124. Joseph Y. S. Cheng, "A Chinese View of China's Energy Security," *Journal of Contemporary China* 17, no. 55 (2008): 297–317.

125. Barton W. Marcois and Leland R. Miller, "China, U.S. Interests Conflict," *Washington Times*, March 24, 2005, accessed July 26, 2012, http://www.washingtontimes.com/news/2005/mar/24/20050324-075950-4488r/.

126. Cheng, "Chinese View of China's Energy Security," 297–317.

127. Cf. Yong Deng, *China's Struggle for Status the Realignment of International Relations* (Cambridge: Cambridge University Press, 2008).

128. "Japanese Activists Land on Disputed Islands," *The Guardian*, August 19, 2012, accessed June 1, 2015, https://www.theguardian.com/world/2012/aug/19/japanese-activists-land-senkaku-islands.

129. Dexter Roberts, "Anti-Japanese Protests Flare in China over Disputed Islands," *Bloomberg.com*, September 17, 2012, accessed September 18, 2012, http://www.bloomberg.com/news/articles/2012-09-17/anti-japanese-protests-flare-in-china-over-disputed-islands.

130. "China-Japan Protests Resume amid Islands Row," *BBC News*, September 18, 2012, accessed September 18, 2012, http://www.bbc.com/news/world-asia-china-19632042.

131. Justin McCurry, "Japan Stokes Tensions with China over Plan to Buy Disputed Islands," *The Guardian*, September 5, 2012, accessed September 18, 2012, https://www.theguardian.com/world/2012/sep/05/japan-china-disputed-islands.

132. Roberts, "Anti-Japanese Protests Flare in China."

133. Weiyi Lim, "China's Stocks Fall on Concern Japan Tensions May Hurt Economy," *Businessweek*, September 18, accessed September 18, 2012, http://www.businessweek.com/news/2012-09-17/china-stock-index-futures-decline-on-pboc-inflation-report.

134. "Chengdu Galaxy Magnets Co," accessed June 20, 2017, http://www.galaxymagnets.com/english/.

135. Lim, "China's Stocks Fall."

136. Lim, "China's Stocks Fall."

137. Lim, "China's Stocks Fall."

138. Katie Hunt and Vivian Kam, "China: South China Sea Island Building 'Almost Complete,'" *CNN*, June 17, 2015, accessed July 7, 2015, http://www.cnn.com/2015/06/17/asia/china-south-china-sea-land-reclamation/index.html; Carrie Gracie, "South China Sea: David v Goliath as Dispute Goes to Court,"

*BBC News*, July 6, 2015, accessed July 7, 2015, http://www.bbc.com/news/world-asia-china-33406609.

139. "International Court Strikes down China's Territorial Claim," *CNBC*, July 12, 2016, accessed August 30, 2016, http://www.cnbc.com/2016/07/12/tensions-in-south-china-sea-to-persist-even-after-court-ruling.html.

140. Fu Jing and An Baijie, "Ruling 'Null and Void', with No Binding Force," *China Daily*, July 13, 2016, accessed September 10, 2016, http://www.chinadaily.com.cn/world/2016-07/13/content_26064418.htm.

141. Jing and Baijie, "Ruling 'Null and Void.'"

142. "Xi Jinping Meets with President Donald Tusk of the European Council and President Jean-Claude Juncker of the European Commission," Ministry of Foreign Affairs of the People's Republic of China, July 12, 2016, accessed September 10, 2016, http://www.fmprc.gov.cn/mfa_eng/zxxx_662805/t1381403.shtml.

143. Norman P. Aquino and Andreo Calonzo, "Duterte Seeks Arms from China, Ends Joint Patrols with U.S.," *Bloomberg.com*, September 13, 2016, accessed September 16, 2016, http://www.bloomberg.com/news/articles/2016-09-13/duterte-courts-china-for-weapons-ends-joint-patrols-with-u-s.

144. Jesse Johnson, "China, Russia to Carry Out Joint Military Drills in South China Sea," *Japan Times*, July 29, 2016, accessed September 16, 2016, http://www.japantimes.co.jp/news/2016/07/29/asia-pacific/china-russia-carry-joint-military-drills-south-china-sea/.

145. Peter Marton and Tamas Matura, "The 'Voracious Dragon,' the 'Scramble' and the 'Honey Pot': Conceptions of Conflict over Africa's Natural Resources," *Journal of Contemporary African Studies* 29, no. 2 (2011): 155–67.

146. Nye, *Soft Power*.

147. Eisenman, Heginbotham, and Mitchell, *China and the Developing World*, 203.

148. Elizabeth Economy, "China's Imperial President," *Foreign Affairs*, November 6, 2014, https://www.foreignaffairs.com/articles/china/2014-10-20/chinas-imperial-president.

149. Adam Segal, "Chinese Economic Statecraft and the Political Economy of Asian Security," in *China's Rise and the Balance of Influence in Asia*, ed. William Keller and Thomas Rawski (Pittsburgh: University of Pittsburgh Press, 2007), 146–61.

150. Shannon Tiezzi, "China's $1 Trillion Investment Plan: Stimulus or Not?," *The Diplomat*, January 8, 2015, accessed February 10, 2015, http://thediplomat.com/2015/01/chinas-1-trillion-investment-plan-stimulus-or-not/; Shannon Tiezzi, "China's 'New Silk Road' Vision Revealed," *The Diplomat*, May 9, 2014, accessed June 12, 2015, http://thediplomat.com//2014/05/chinas-new-silk-road-vision-revealed/; "Gov't Said to Name Three to Silk Road Fund Leadership Team," *Caixin Online*, May 5, 2015, accessed September 25, 2016, http://english.caixin.com/2015-02-05/100781902.html.

151. Masahiro Okoshi, "Broader Appeal: European Countries Plan to Join China-Led Infrastructure Bank," *Nikkei Asian Review*, March 7, 2015, accessed September 25, 2016, http://asia.nikkei.com/Politics-Economy/Policy-Politics/European-countries-plan-to-join-China-led-infrastructure-bank; "Why China Is Creating a New 'World Bank' for Asia," *The Economist*, November 11, 2014, accessed

December 10, 2014, http://www.economist.com/blogs/economist-explains/2014/11/economist-explains-6.

152. "President's Opening Statement 2016 Annual Meeting of the Board of Governors Asian Infrastructure Investment Bank," Asian Infrastructure Investment Bank (AIIB), Beijing, June 25, 2016, accessed September 25, 2016, https://www.aiib.org/en/news-events/news/2016/20160625_001.html.

153. Gracie, "South China Sea: David v Goliath."

154. Segal, "Chinese Economic Statecraft."

155. Vincent Wei-cheng Wang, "China's Economic Statecraft toward Southeast Asia: Free Trade Agreement and 'Peaceful Rise,'" *American Journal of Chinese Studies* 13, no. 1 (2006): 5–34. Cf. Yong Deng, *China Rising: Power and Motivation in Chinese Foreign Policy* (Lanham, MD: Rowman and Littlefield, 2005).

156. Neil Thompson, "China's Growing Presence in Russia's Backyard," *The Diplomat*, March 25, 2015, accessed September 25, 2016, http://thediplomat.com/2015/03/chinas-growing-presence-in-russias-backyard/.

## CHAPTER 2

1. Gordon B. Haxel, James B. Hedrick, and Greta J. Orris, "Rare Earth Elements: Critical Resources for High Technology" (Fact Sheet 087-02, U.S. Geological Survey, Washington, DC), accessed January 2, 2013, http://pubs.usgs.gov/fs/2002/fs087-02/.

2. US Geological Survey, "2010 Minerals Yearbook: Rare Earths" (Washington, DC: US Geological Survey, September 2012), accessed December 10, 2012, http://minerals.usgs.gov/minerals/pubs/commodity/rare_earths/myb1-2010-raree.pdf.

3. James B. Hendrick, "Rare-Earth Metal Prices in the USA ca. 1960–1994," *Journal of Alloys and Compounds* 250 (1997): 471.

4. The special chemical and physical properties are well-documented and thoroughly described in scientific encyclopedias and in the chemistry and physics literature. See "ROEMPP—Thieme Chemistry—Georg Thieme Verlag," *Thieme*, accessed September 26, 2016, https://www.thieme.de/en/thieme-chemistry/roempp-54669.htm; Charles E. Mortimer, *Chemie: das Basiswissen der Chemie; 123 Tabellen* (Stuttgard: Thieme, 2001). Cf. Volker Zepf, "Rare Earth Elements: A New Approach to the Nexus of Supply, Demand and Use: Exemplified along the Use of Neodymium in Permanent Magnets" (PhD diss., University of Augsburg, 2013).

5. "Rare Earths Statistics and Information," US Geological Survey, accessed January 10, 2013, http://minerals.usgs.gov/minerals/pubs/commodity/rare_earths/.

6. USGS, "2010 Minerals Yearbook: Rare Earths."

7. USGS, "2010 Minerals Yearbook: Rare Earths"; John Seaman, "Rare Earths and Clean Energy: Analyzing China's Upper Hand" (Paris, Institut français des relations internationales, September 2010), accessed May 10, 2011, www.Ifri.org.

8. Hurst, "China's Rare Earth Elements Industry."

9. C. K. Gupta and N. Krishnamurthy, *Extractive Metallurgy of Rare Earths* (Boca Raton, Florida: CRC Press, 2015), 88.

10. Roskill Information Services, *The Economics of Rare Earths and Yttrium* (London: Roskill Information Services, 2007); Zepf, "Rare Earth Elements."

11.  "Mineral Commodity Summaries" (USGS, January 2015), 128, accessed June 19, 2016, https://minerals.usgs.gov/minerals/pubs/mcs/2015/mcs2015.pdf.

12.  Incandescent mantles were used in Europe and North America for lighting in the late nineteenth century.

13.  James B. Hedrick, "Thorium," USGS.gov, accessed January 28, 2013, minerals.usgs.gov/minerals/pubs/commodity/thorium/690494.pdf.

14.  Artem Golev et al., "Rare Earths Supply Chains: Current Statues, Constraints and Opportunities," *Resources Policy* 41 (2014): 52–59.

15.  Niluksi Koswanage, "UPDATE 2-Malaysia Approves Temporary License for Lynas Rare Earths Plant," *Reuters*, February 1, 2012, accessed May 10, 2012, http://www.reuters.com/article/australia-lynas-malaysia-idUSL4E8D15KQ20120201; "Australia's Lynas Rare Earth Plant Made History in Malaysia Today: Nationwide Rally to Stop the Project," Stoplynus.org, website of the Australian Stop Lynus campaign, n.d., accessed February 17, 2013, http://stoplynas.org/australia%E2%80%99s-lynas-rare-earth-plant-made-history-in-malaysia-today-nationwide-rally-to-stop-the-project/.

16.  "Lynas Gets New License for Kuantan Rare-Earth Plant as TOL Expires," *Malayi Mail Online*, September 2, 2014, accessed December 10, 2014, http://www.themalaymailonline.com/malaysia/article/lynas-gets-new-licence-for-kuantan-rare-earth-plant-as-tol-expires.

17.  World Nuclear Association, "Thorium," World Nuclear Association.org, Information Library," updated February 2017, accessed March, 2017, http://www.world-nuclear.org/information-library/current-and-future-generation/thorium.aspx.

18.  Nicholas Jepson, "A 21st Century Scramble: South Africa, China and the Rare Earth Metals Industry," (SAIIA occasional paper no. 113, South African Institute of International Affairs, Johannesburg, SA, March 2012), January 28, 2013, http://www.saiia.org.za/images/stories/pubs/occasional_papers_above_100/saia_sop_113_jepson_20120315.pdf.

19.  Molycorp filed for bankruptcy in 2015 and, in August 2015, announced that it would "transition its Mountain Pass Rare Earth Facility to a 'care and maintenance' mode while it plans to continue serving its rare earth oxide customers via its production facilities in Estonia and China." See "Molycorp to Move Its Mountain Pass Rare Earth Facility to 'Care and Maintenance' Mode," Molycorp.com, accessed August 30, 2015, http://www.molycorp.com/molycorp-to-move-its-mountain- pass-rare-earth-facility-to-care-and-maintenance-mode/; "Molycorp, Inc. Files Restructuring Plan," *Molycorp*, November 3, 2015, accessed December 20, 2015, http://www.molycorp.com/molycorp-inc-files-restructuring-plan/.

20.  "Rare Earth Elements: Critical Resources for High Technology" (Fact Sheet 087-02, USGS), accessed January 2, 2013, http://pubs.usgs.gov/fs/2002/fs087-02/.

21.  "Rare Earths," US Geological Survey, January 2016, accessed September 14, 2016, http://minerals.usgs.gov/minerals/pubs/commodity/rare_earths/mcs-2016-raree.pdf.

22.  Jepson, "SAIIA: A 21st Century Scramble."

23.  Nagaiyar Krishnamurthy and Chiranjib Kumar Gupta, *Extractive Metallurgy of Rare Earths*, 2nd ed. (Boca Raton, FL: CRC Press, 2015), https://www.crcpress.

com/Extractive-Metallurgy-of-Rare-Earths-Second-Edition/Krishnamurthy-Gupta/p/book/9781466576346; K. A. Gschneidner, L. Eyring, and M. B. Maple, *Handbook on the Physics and Chemistry of Rare Earths: High Temperature Rare Earths Superconductors—I* (Amsterdam: Elsevier, 2000).

24. Dr. Steve Ward, "SACOME Breakfast Briefing, September 2011," Arafura Resources, accessed August 20, 2012, http://www.asx.com.au/asxpdf/20110915/pdf/421364tmvn07qh.pdf.

25. *"Technology Metals Research LLC,"* updated August 2011, accessed January 3, 2012, www.techmetalsresearch.com. For more detailed references about the properties of rare earths, see Jacques Lucas et al., *Rare Earths: Science, Technology, Production and Use,* 1 edition (Amsterdam: Elsevier, 2014).

26. U. S. Environmental Protection Agency (EPA), "Rare Earth Elements: A Review of Production, Processing, Recycling, and Associated Environmental Issues " (report EPA/600/R-12/572, US EPA, Washington, DC, December 2012) accessed August 10, 2016, http://reviewboard.ca/upload/project_document/EA1011-001_US_EPA_-_Rare_Earth_Elements_-_Associated_Environmental_Issues.PDF. For a more detailed description, see Krishnamurthy and Gupta, *Extractive Metallurgy of Rare Earths.*

27. US EPA, "Rare Earth Elements."

28. Hurst, "China's Rare Earth Elements Industry?"

29. Hurst, "China's Rare Earth Elements Industry?"

30. Hurst, "China's Rare Earth Elements Industry?"

31. US EPA, "Rare Earths."

32. US EPA, "Rare Earths."

33. Neale R. Neelameggham et al., *Rare Metal Technology 2014* (Hoboken, New Jersey: John Wiley & Sons, 2014), 82–83.

34. Francois Cardarelli, *Materials Handbook: A Concise Desktop Reference* (London: Springer, 2008), 427.

35. Liz Nickels, "The Growing Pull of Rare Earth Magnets," *Metal Powder Report* 65, no. 2 (2010): 6–8.

36. Nickels, "Growing Pull of rare earth magnets," 6–8.

37. Reza Sadeghbeigi, *Fluid Catalytic Cracking Handbook: An Expert Guide to the Practical Operation, Design, and Optimization of FCC Units* (Amsterdam: Elsevier, 2012).

38. US Department of Energy, "Critical Materials Strategy," December 2011, accessed May 10, 2012, http://energy.gov/sites/prod/files/DOE_CMS2011_FINAL_Full.pdf.

39. Sadeghbeigi, *Fluid Catalytic Cracking Handbook*; M. Kostoglou, A. G. Konstandopoulos, and H. Burtscher, "Size Distribution Dynamics of Fuel-Borne Catalytic Ceria Nanoparticle," *Journal of Aerosol Science* 38, no. 6 (2007): 604–11; S. Lorentzou, C. Pagkoura, A. Zygogianni, G. Kastrinaki, A. G. Kostandopoulos, 'Catalytic Nano-Structured Materials for Next Generation Diesel Particulate Filters," *SAE International Journal of Materials and Manufacturing* 1 (2009): 181–98.

40. See Jacques Lucas et al., *Rare Earths: Science, Technology, Production and Use,* 1 edition (Amsterdam: Elsevier, 2014).

41. R. L. Moss, E. Tzimas, H. Kara, P. Willis, and J. Kooroshy, "The Potential Risks from Metals Bottlenecks to the Deployment of Strategic Energy Technologies,"

*Energy Policy* 55 (2013): 556–64; Stefania Massari and Marcello Ruberti, "Rare Earth Elements as Critical Raw Materials: Focus on International Markets and Future Strategies," *Resources Policy* 38, no. 1 (2013): 36–43.

42. US Department of Energy, "Critical Materials Strategy."
43. "Rare-Earth Separation," Innovation Metals Corp, accessed September 26, 2016, http://www.innovationmetals.com/rare-earth-separation/.
44. US Department of Energy, "Critical Materials Strategy."
45. Keith Bradsher, "In China, Gang-Run Illegal Rare Earth Mines Face a Crackdown," *New York Times*, December 29, 2010, accessed January 28, 2013, http://www.nytimes.com/2010/12/30/business/global/30smuggle.html.
46. US Government Accountability Office (GAO), "Rare Earths Materials in the Defense Supply Chain: Briefing for Congressional Committees," April 1, 2010, accessed February 2, 2012, http://www.gao.gov/assets/100/96654.pdf.
47. GAO, "Rare Earths Materials in the Defense Supply Chain
48. GAO, "Rare Earths Materials in the Defense Supply Chain."
49. UN Framework Convention on Climate Change (UNFCCC), "The Paris Agreement," accessed October 9, 2016, http://unfccc.int/paris_agreement/items/9485.php.
50. There were 28 EU member states as of 2013, but will be 27 again after the completion of the Brexit.
51. José Manuel Durão Barroso, "Thinking like Scientists, Acting like Leaders" (press release, European Commission, June 16, 2011), accessed January 29, 2013, http://europa.eu/rapid/press-release_SPEECH-11-441_en.htm.
52. US Department of Energy, "Critical Materials Strategy."
53. Sanjeev Miglani, "With an Eye on China, India Steps Up Defense Spending," *Reuters*, February 28, 2011, accessed January 15, 2012, http://www.reuters.com/article/2011/02/28/india-budget-military-idUSSGE71R02Y20110228; Sanjeev Miglani, "India Raises Military Spending, Eases Foreign Investment Limit in Arms Industry," *Reuters*, July 10, 2014, accessed December 1, 2015, http://in.reuters.com/article/india-budget-defence-idINKBN0FF0WQ20140710.
54. Richard A. Bitzinger, "China's Double-Digit Defense Growth: What It Means for a Peaceful Rise," *Foreign Affairs*, March 19, 2015, accessed December 1, 2015, https://www.foreignaffairs.com/articles/china/2015-03-19/chinas-double-digit-defense-growth; Ben Blanchard, "China Defense Budget to Stir Regional Disquiet," *Reuters*, Mar 4, 2011, accessed November 2011, http://www.reuters.com/article/2011/03/04/us-china-defence- idUSTRE7230ZN20110304.
55. David Watt, "Defense Budget Overview," Parliament of Australia, accessed December 1, 2015, http://www.aph.gov.au/About_Parliament/Parliamentary_Departments/Parliamentary_Library/pubs/rp/BudgetReview201415/DefenceBudget; Ian McPhedran,"Weary Warriors of the West Want to Sheath Swords," *Daily Telegraph*, January 10, 2012, accessed on March 15, 2012, http://www.dailytelegraph.com.au/news/opinion/weary-warriors- of-the-west-want-to-sheath-swords/story-e6frezz0-1226240223665.
56. "Ames Laboratory to Lead New Research Effort to Address Shortages in Rare Earth and Other Critical Materials," *US Department of Energy*, January 9, 2013,

accessed January 27, 2013, http://energy.gov/articles/ames-laboratory-lead-new-research-effort-address-shortages-rare-earth-and-other-critical.

57. "Ames Laboratory to Lead New Research Effort to Address Shortages in Rare Earth and Other Critical Materials."

58. "Ames Laboratory to Lead New Research Effort to Address Shortages in Rare Earth and Other Critical Materials."

59. "A rare opportunity," *Metal Powder Report* 48, Issue 2 (February 1993): 18–19.

60. Yang Luo, "The Inexorable Rise of China's NdFeB Magnet Industry," *Metal Powder Report* 63, no. 11 (December 2008): 8–10.

61. Yang Luo, "The Inexorable Rise of China's NdFeB Magnet Industry," 8–10.

62. Steven Constantinides, "Market Outlook for Ferrite, Rare Earth and Other Permanent Magnets," January 21, 2016, accessed September 30, 2016, http://www.magneticsmagazine.com/conferences/2016/Presentations/Arnold_Constantinides.pdf.

63. Seaman, "Rare Earths and Clean Energy."

64. "Renewables 2016 Global Status Report," *REN21*, 2016, accessed September 20, 2016, http://www.ren21.net/wp-content/uploads/2016/06/GSR_2016_Full_Report.pdf.

65. US Department of Energy, "Critical Materials Strategy."

66. US Department of Energy, "Critical Materials Strategy"; According to Henrik Stiesdal, CTO of Siemens Wind Power:

    Intended for higher power output, including offshore applications, Direct drive generators rely on rare-earth PM materials—typically neodymium iron boron (Nd-Fe-B). However, technology offers no free lunches. A trade-off for eliminating the gearbox is the need for large quantities of these costly magnet materials, which are becoming subject to supply shortages. Approximately 650 kg of PMs is needed per MW wind turbine capacity, according to Siemens, of which 25%-30% is rare-earth magnet material. The cost of these materials is accounted for in the competitive price of the direct-drive concept . . . Generators currently offered use permanent magnets with neodymium and dysprosium elements. (Frank J. Bartos PE, "Direct-Drive Wind Turbines Flex Muscles," *Control Engineering*, July 15, 2011, January 28, 2013, http://www.controleng.com/single-article/direct-drive-wind-turbines-flex-muscles/4be132ffb053a53fc4ab10b2c9c57340.html)

67. "Renewables 2016 Global Status Report."

68. "Renewables 2016 Global Status Report."

69. "Phosphor:—A Critical Component in Fluorescent Lamps," Philips, accessed January 28, 2013, http://www.usa.lighting.philips.com/pwc_li/us_en/lightcommunity/trends/phosphor/assets/Philips_REO_Brochure_P-6281.pdf; "Customer Information on Rare Earths," OSRAM, accessed January 28, 2013, http://www.osram.com/media/resource/HIRES/333009/264096/pdf_osramcustomer_information_rare_earth_metals_eng.pdf.

70. "FAQ: Phasing Out Conventional Incandescent Bulbs," (MEMO/09/368), September 1, 2009, accessed January 28, 2013, http://europa.eu/rapid/pressReleasesAction.do?reference=MEMO/09/368.

71. "EU to Phase Out Standard Light Bulbs by 2012," *Associated Press*, December 8, 2008, accessed June 12, 2012, http://www.foxnews.com/story/0,2933,463664,00.

html. Recently the EU voted to close loopholes that allowed appliance manufactures to make ambitious and misleading claims about their product's energy performance but this has not been extended to the lighting industry. It is reported that the Commission fears that stricter rules may have serious impacts on the lighting industry. This is causing heated debate and reactions on the part of environmentalists. See Arthur Neslen, "Lightbulbs Excluded in EU Regulations on Energy Efficiency Claims," *The Guardian*, April 12, 2016, sec. Environment, https://www.theguardian.com/environment/2016/apr/12/lightbulbs-lighting-excluded-eu-regulations-energy-efficiency-advertising.

72. "FAQ: Phasing Out Conventional Incandescent Bulbs," (MEMO/09/368, September 1, 2009, accessed on January 28, 2013, http://europa.eu/rapid/pressReleasesAction.do?reference=MEMO/09/368.

73. "Shedding New Light on the U.S. Energy Efficiency Standards for Everyday Light Bulbs," National Resources Defense Council, January 2013, accessed on January 28, 2013, http://www.nrdc.org/energy/energyefficientlightbulbs/files/shedding-new-light-FS.pdf; "New Lighting Standards Began in 2012," Energy.gov, accessed September 1, 2016, http://energy.gov/energysaver/new-lighting-standards-began-2012.

74. Julian M. Allwood et al., "Material Efficiency: Providing Material Services with Less Material Production," *Philosophical Transactions of the Royal Society A* 371, no. 1986 (March 2013).

75. Robert A. Frosch and Nicholas E. Gallopoulos, "Strategies for Manufacturing," *Scientific American* 261, no. 3 (1989): 144–52.

76. T. E. H. Graedel and Braden R. Allenby, *Industrial Ecology* (Englewood Cliffs, NJ: Prentice Hall, 1995), 9.

77. Mark Halper, "Japanese manufacturers to China: We don't need your rare earths," *Smart Planet*, October 12, 2012, accessed January 27, 2013, http://www.smartplanet.com/blog/bulletin/japanese-%20manufacturers-to-china-we-dont-need-your-rare-earths/2316; "Companies Seek to Reduce Dependence on Chinese REEs," *MINING.com*, November 10, 2013, http://www.mining.com/web/companies-seek-to-reduce-dependence-on-chinese-rees/.

78. US Department of Energy, "Critical Materials Strategy."

79. Guochang Xu, Junya Yano, and Shin-ichi Sakai, "Scenario Analysis for Recovery of Rare Earth Elements from End-of-Life Vehicles," *Journal of Material Cycles and Waste Management* 18, no. 3 (2016): 469–82, doi:10.1007/s10163-016-0487-y.

80. Elisa Alonso et al., "Evaluating Rare Earth Element Availability: A Case with Revolutionary Demand from Clean Technologies," *Environmental Science and Technology* 46, no. 6 (2012): 3406–14, doi:10.1021/es203518d; Koen Binnemans et al., "Recycling of Rare Earths: A Critical Review," *Journal of Cleaner Production* 51 (July 2013): 1–22, doi:10.1016/j.jclepro.2012.12.037.

81. Alonso, "Evaluating Rare Earth Element Availability."

82. Xu, Yano, and Sakai, "Scenario Analysis for Recovery of Rare Earth Elements," 469–82.

83. Achilleas Tsamis and Mike Coyne, "Recovery of Rare Earths from Electronic Wastes: An Opportunity for High-Tech SMEs," *Directorate General for Internal Policies: Policy Department A: Economic and Scientific Policy for the European*

*Parliament*, February 2015, accessed October 10, 2016, http://www.europarl.europa.eu/RegData/etudes/STUD/2015/518777/IPOL_STU(2015)518777_EN.pdf

84. US Department of Energy, "Critical Materials Strategy."

85. Golev, "Rare Earths Supply Chains," 52–59.

86. Doris Schüler, Matthias Buchert, et al., "Study on Rare Earths and Their Recycling," Öko-Institut, January 2011, accessed December 20, 2013, http://www.oeko.de/oekodoc/1112/2011-003-en.pdf.

87. Cindy Hurst, "The Rare Earth Dilemma: China's Rare Earth Environmental and Safety Nightmare," *Cutting Edge*, November 15, 2010, accessed January 28, 2013, http://www.thecuttingedgenews.com/index.php?article=21777.

88. Rare earths are but one resource in a long list of others whose production has had social and environmental impacts on the poor. For a wider discussion on energy production and social marginalization in China, Cf. Philip Andrews-Speed and Xin Ma, "Energy Production and Social Marginalisation in China," *Journal of Contemporary China* 17, no. 55 (2008): 247–72.

89. "Rare Earths: Shades of Grey," *China Water Risk*, June 2016, accessed September 20, 2016, http://chinawaterrisk.org/wp-content/uploads/2016/08/China-Water-Risk-Report-Rare-Earths-Shades-Of-Grey-2016-Eng.pdf.

90. "Ganzhou Needs over 3 Billion Yuan for Environment Restoration of Rare Earth Mineral Mines," accessed October 9, 2016, http://www.chinaenvironment.info/EcoSystems/201606/t20160615_67668.html.

91. Hurst, "Rare Earth Dilemma."

92. Hurst, "Rare Earth Dilemma."

93. "Situation and Policies of China's Rare Earth Industry," June 2012, accessed December 10, 2012, 2016, http://english.gov.cn/archive/white_paper/2014/08/23/content_281474983043156.htm.

## CHAPTER 3

1. Robert P. Multhauf, *Neptune's Gift: A History of Common Salt* (Baltimore: Johns Hopkins University Press, 1978), 3.

2. Jean-François Bergier, *Une Histoire du sel* (Fribourg: Presses Universitaires de France, 1982), 11.

3. Bergier, *Histoire du sel*, 13 (translated from the French by the author).

4. Garnett Eskew, *Salt, The Fifth Element: The Story of a Basic American Industry.* (Chicago: J. G. Ferguson, 1948), 12.

5. Menashe Har-El, "The Routes of Salt, Sugar and Balsam Caravans in the Judean Dessert," *GeoJournal* 2, no. 6 (1978): 549–56.

6. Harlan W. Gilmore, "Cultural Diffusion via Salt," *American Anthropologist* 57, no. 5 (1955): 1011–15.

7. Multhauf, *Neptune's Gift*, 4.

8. Rowan K. Flad, *Salt Production and Social Hierarchy in Ancient China: An Archaeological Investigation of Specialization in China's Three Gorges* (Cambridge: New York: Cambridge University Press, 2011).

9. Flad, *Salt Production and Social Hierarchy*, 37.

10. Multhauf, *Neptune's Gift*, 12.

11. Cf. Friedrich Hirth, *The Ancient History of China* (New York: Columbia University Press, 1908); Michael Loewe and Edward L. Shaughnessy, *The Cambridge History of Ancient China: From the Origins of Civilization to 221 BC* (New York: Cambridge University Press, 1999); John S. Major, *Ancient China: A History* (New York: Routledge, 2016).

12. Flad, *Salt Production and Social Hierarchy*, 35.

13. Esson M. Gale, "Public Administration of Salt in China: A Historical Survey," *Annals of the American Academy of Political and Social Science* 152 (November 1930): 241–51.

14. Mark Kurlansky, *Salt: A World History* (New York: Penguin, 2003), 31.

15. Joseph Earle Spencer, "Salt in China," *Geographical Review* 25, no. 3 (1935): 353–66.

16. Huan K'uan, *Discourses on Salt and Iron: A Debate on State Control of Commerce and Industry in Ancient China, Chapters I–XIX* (Leiden: E. J. Brill, 1931), 33.

17. Huan, *Discourses on Salt and Iron*, 3–5.

18. Huan, *Discourses on Salt and Iron*, 3–5.

19. Kurlansky, *Salt: A World History*, 29.

20. E. H. Parker, "The Chinese Salt Trade: An Opening for British Enterprise," *Economic Journal*, vol. 10, 1900.

21. Tao-Chang Chiang, "The Salt Trade in Ch'ing China," *Modern Asian Studies* 17, no. 2 (1983): 218. The continued and extensive smuggling of rare earths is one the key issues that the PRC government has been trying to address, especially since 2010.

22. Flad, *Salt Production and Social Hierarchy*, 35.

23. Kurlansky, *Salt: A World History*, 26.

24. Kurlansky, *Salt: A World History*, 28.

25. Zbigniew Brzeszinski, *The Grand Chessboard* (New York: Basic, 1997), 16. Today, China is increasingly looking to use rare earths in applications for domestic consumption.

26. S. A. M Adshead, *Salt and Civilization* (New York: St. Martin's, 1991), 4.

27. Adshead, *Salt and Civilization*, 10.

28. Adshead, *Salt and Civilization*, 33.

29. Kurlansky, *Salt: A World History*, 65.

30. Multhauf, *Neptune's Gift*, 8.

31. Kurlansky, *Salt: A World History*, 84.

32. Jean-Claude Hocquet, "Capitalisme marchand et classe marchande à Venise au temps de la Renaissance," *Annales. Économies, Sociétés, Civilisations* 34ᵉ année, n. 2 (1979): 279–304.

33. Hocquet, "Capitalisme marchand et classe marchande," 279–304.

34. Bergier, *Histoire du sel*, 137.

35. Kurlansky, *Salt: A World History*, 120.

36. Multhauf, *Neptune's Gift*, 9.

37. Eskew, *Salt, The Fifth Element*, 17.

38. Bergier, *Histoire du sel*, 204.

39. Bergier, *Histoire du sel*, 138.

40. Kurlansky, *Salt: A World History*, 200–213.

41. Bergier, *Histoire du sel*, 138.

42. Ella Lonn, *Salt as a Factor in the Confederacy* (University: University of Alabama Press, 1965), 15.

43. Bergier, *Histoire du sel*, 36.

44. Kurlansky, *Salt: A World History*, 257.

45. Lonn, *Salt as a Factor in the Confederacy*, 13.

46. Yogesh Chadha, *Gandhi: A Life* (New York: John Wiley, 1997), 287–89.

47. Thomas K. Weber, *On the Salt March: The Historiography of Mahatma Gandhi's March to Dandi*, 2nd rev. ed. (New Delhi: Rupa & Co., 2009).

48. Chadha, *Gandhi: A Life*, 294.

49. Kurlansky, *Salt: A World History*, 352–53.

50. Kurlansky, *Salt: A World History*, 297.

51. Nicolas Appert, *L'art de conserver pendant plusieurs années toutes les substances animales et végétales* (Paris: Patris et Cie, 1810).

52. Appert, *L'art de conserver*.

53. Mark Kurlansky, *Birdseye: The Adventures of a Curious Man* (New York: Anchor, 2012).

54. Kurlansky, *Birdseye: The Adventures of a Curious Man*.

55. Kurlansky, *Salt: A World History*, 303–8.

56. A popular poem described the discontent over the French gabelle:

    I mourn the salt of Provence,
    Because at my tollgate nothing
    Has passed through. (The Count)
    made the salt so expensive
    That I fear more and more
    That the popular proverb
    Might unfortunately come true
    The transport of meat is completely lost without salt
    But the blame is with the Lord (Charles d'Anjou)
    Who discourages his own . . . (Bergier, *Histoire du sel*, 205; my translation)

57. Bergier, *Histoire du sel*, 12.

58. Kurlansky, *Salt: A World History*, 316.

59. "Crude Oil," NASDAQ.com, accessed September 12, 2016, http://www.nasdaq.com/markets/crude-oil.aspx?timeframe=1y; "Short-Term Energy Outlook: Crude Oil Prices" (forecast, US Energy Information Administration, Independent Statistics and Analysis, Washington, DC, September 2016), accessed September 10, 2016, https://www.eia.gov/forecasts/steo/report/prices.cfm.

60. In addition, the ongoing rivalry between Saudi Arabia and Iran has also played a role, especially after most of the sanctions against Iran were lifted.

61. "Aprovlepti I Anodos tou Petrelaiou," *Kathimerini*, March 10, 2015, accessed 10 March 2015, http://www.kathimerini.gr/806695/article/oikonomia/die8nhs-oikonomia/aprovlepth-anodos-toy-petrelaioy.

62. Ahmed Al Omran, Benoit Faucon, and Summer Said, "Saudi Arabia, Russia Agree to New Oil Pact, but No Output Freeze," *Wall Street Journal*, September

5, 2016, accessed September 14, 2016, http://www.wsj.com/articles/saudi-arabia-russia-agree-to-new-oil-pact-but-no-output-freeze-1473072395.

63. China was a net oil exporter until the early 1990s and became the world's second largest net importer of oil in 2009. Today it is the world's largest importer of oil. See "China," U.S. Energy Information Administration, November 2012, accessed November 10, 2012, http://www.eia.gov/countries/cab.cfm?fips=CH.

64. "China." "US net oil imports dropped to 5.98m barrels a day in December, the lowest since February 1992, according to provisional figures from the US Energy Information Administration. In the same month, China's net oil imports surged to 6.12m b/d, according to Chinese customs." Javier Blas, "China Becomes World's Top Oil Importer," *Financial Times*, March 4, 2013, accessed May 20, 2013, http://www.ft.com/intl/cms/s/0/d33b5104-84a1-11e2-aaf1-00144feabdc0.html.

65. Joseph Y. S. Cheng, "A Chinese View of China's Energy Security," *Journal of Contemporary China* 17, no. 55 (2008): 297–317.

66. Chuka Enuka, "The Forum on China-Africa Cooperation (Focac): A Framework for China's Re-engagement with Africa in the 21st Century," *E-BANGI: Journal of Social Sciences and Humanities* 6, no. 2 (2011): 220.

67. Sigfrido Burgos Cáceres and Sophal Ear, "The Geopolitics of China's Global Resources Quest," *Geopolitics* 17, no. 1 (2012): 47–79; Joshua Eisenman, Eric Heginbotham, and Derek Mitchell, *China and the Developing World: Beijing's Strategy for the Twenty-First Century* (Armonk, NY: M. E. Sharpe, 2007).

68. John Lee, "The 'Tragedy' of China's Energy Policy," *The Diplomat* blogs, October 4, 2012, accessed November 1, 2012, http://thediplomat.com/china-power/the-tragedy-of-chinas-energy-policy/. According to Lee, the CNOOC (China National Offshore Oil Corporation) offered $15.1 billion for Nexen Inc., a Canadian energy extraction firm that owns assets in Alberta, the Gulf of Mexico, and the North Sea. The amount was unprecedented because the company was valued at $6.5 billion, and the move gave rise to fears that China was again acting strategically vis-à-vis the securing of energy supplies, especially oil.

69. As Cao and Bluth describe it in their article pertaining to China's energy security, "It (energy) is also a strategic material and major element of a country's security, which links national and foreign security policies. Energy is becoming more and more important along with competition of world energy market and rising concerns about energy security. Energy security is a systemic issue involving engineering, politics, economics, energy, environment, diplomacy and military affairs. Given the extensive control the state exercises over the economy, the management of energy resources is crucial and requires China's government to play a much more intrusive and pro-active role in this process than is the case in other major economies." Cao and Bluth, "Challenges and Countermeasures of China's Energy Security," 381–88.

70. China had made clear its intention to invest in renewables both in its eleventh and twelfth five-year plans. "Abstract of the 11th Five Year Plan," Central People's Government of the People's Republic of China (English-language) website, English. gov.cn, March 8, 2006, accessed January 25, 2013, http://www.gov.cn/english/special/115y_index.htm; Joseph Casey, Katherine Koleski, "Backgrounder: China's 12th

Five Year Plan" (report of US-China Economic and Security Review Commission, Washington, DC, June 24, 2011), accessed January 25, 2013, https://www.uscc.gov/Research/backgrounder-china's-12th-five-year-plan; "China's 12th Five Year Plan," APCO worldwide, December 10, 2010, accessed January 25, 2013, http://apco-worldwide.com/content/PDFs/Chinas_12th_Five-Year_Plan.pdf; "China's 12th Five Year Plan," APCO worldwide, December 10, 2010, accessed June 20, 2017, https://sustainabledevelopment.un.org/index.php?page=view&type=400&nr=70 0&menu=1515.

71.  In the 1950s, M. King Hubbert, a geophysicist who worked for Shell Oil, claimed that oil from US wells would peak around 1970 and decline rapidly after that, and that oil reserves were finite and our rates of consumption would eventually out-pace the discovery of new reserves. Hubbert's predictions were proven correct, though at the time his peers did not share his so-called pessimistic prediction. Hubbert's techniques have been used to evaluate the world supply of oil and have produced similar results indicating a swiftly approaching peak point followed by rapid decline.

72.  Catherine Gautier, *Oil, Water, and Climate: An Introduction* (New York: Cambridge University Press, 2008), 82; David Goodstein, *Out of Gas: The End of the Age of Oil* (New York: Norton, 2005), 17.

73.  Peter R. Odell, "Towards a Geographically Reshaped World Oil Industry," *World Today* 37, no. 12 (1981): 447–53.

74.  Robert E. Ebel, "The Geopolitics of Energy into the 21st Century," U.S. Department of State, April 30, 2002, accessed March 10, 2012, https://2001-2009.state.gov/s/p/of/proc/tr/10187.htm.

75.  Leonardo Maugeri, *The Age of Oil: The Mythology, History, and Future of the World's Most Controversial Resource* (Westport, CT: Praeger, 2006).

76.  Gad G. Gilbar, *The Middle East Oil Decade and Beyond: Essays in Political Economy* (London: Frank Cass, 1997), 17.

77.  OPEC, "Member Countries," *OPEC.com*, accessed January 1, 2013, http://www.opec.org/opec_web/en/about_us/25.htm.

78.  Anne Korin and Gal Luft, *Turning Oil into Salt: Energy Independence through Fuel Choice* (Charleston, SC: BookSurge, 2009), 20.

79.  Brai Odion-Esene, "Opec Oil Revenue to Breach $1 Tln Mark," *Forexlive*, June 14, 2011, accessed July 5, 2012, http://www.forexlive.com/blog/2011/06/14/us-eia-projects-2011-opec-oil-revenue-to-breach-1-tln-mark/. This has yet to materialize because of the fluctuating oil prices of the last two years. Revenues stood at $824 billion in 2013 and fell to $730 billion in 2014. "OPEC Revenues Fact Sheet," US Energy Information Administration, March 31, 2015, viewed September 20, 2015, https://www.eia.gov/beta/international/regions-topics.cfm?RegionTopicID=OPEC.

80.  Roger Stern, "Oil Market Power and United States National Security," *PNAS* 103, no. 5 (2006): 1650–55.

81.  Nouriel Roubini and Brad Setser, "The Effects of the Recent Oil Price Shock on the U.S. and Global Economy," August 2004, accessed October 15, 2012, http://pages.stern.nyu.edu/~nroubini/papers/OilShockRoubiniSetser.pdf.

82. Roubini goes on to say that in the particular time, "other factors were more important: the bust of the internet bubble, the collapse of real investment and, in smaller measure, the Fed tightening between 1999 and 2000." Roubini and Setser, "Effects of the Recent Oil Price Shock."

83. Daniel Yergin, *The Prize* (New York: Free Press, 2008), 595–614.

84. Robert J. Lieber, *The Oil Decade: Conflict and Cooperation in the West* (New York: Praeger, 1983), 3.

85. Robert O. Keohane, *After Hegemony: Cooperation and Discord in the World Political Economy* (Princeton, NJ: Princeton University Press, 1984), 222–23.

86. "How Does the IEA Respond to Major Disruptions in Oil Supply?," International Energy Agency, 2017, March 10, 2011, accessed June 20, 2017, https://www.iea.org/newsroom/news/2011/march/how-does-the-iea-respond-to-major--disruptions-in-the-supply-of-oil-2011-03-10-.html.

87. Keohane, *After Hegemony*, 223.

88. Keohane, *After Hegemony*, 223.

89. "How Does the IEA Respond?"

90. Crude Oil and Commodity Prices, April 2, 2012, accessed April 2, 2012, http://www.oil-price.net/.

91. The EU Accounts for the Greatest Share of World Trade and Generates One Quarter of Global Wealth. See *The EU in the World: The Foreign Policy of the European Union," European Union* (report by European Commission, Brussels, June 2007), accessed January 20, 2013, http://ec.europa.eu/archives/publications/booklets/move/67/en.pdf.

92. Stern, "Oil Market Power."

93. Steven Mufson, "As China, U.S. Vie for More Oil, Diplomatic Friction May Follow," *Washington Post*, April 16, 2006, accessed December 2, 2012, http://www.washingtonpost.com/wp-dyn/content/article/2006/04/14/AR2006041401682.html.

94. Iraq invaded Kuwait and occupied it for seven months. This action on the part of Iraq led directly to the first Iraq war with the United States and its allies. There are a number of reasons Saddam Hussein may have played such an aggressive hand. First, Iraq accused Kuwait of stealing oil by slant drilling. Second, Iraq owed Kuwait over $80 billion, which it had borrowed in order to conduct the war with Iran. Third, Kuwait's overproduction was driving oil prices down and thus diminishing Iraq's revenues.

95. Michael L. Ross, "Blood Barrels: Why Oil Wealth Fuels Conflict," *Foreign Affairs* (May–June 2008): 2–9.

96. Ross, "Blood Barrels."

97. China GDP Annual Growth Rate, *Trading Economics*, accessed September 28, 2016, http://www.tradingeconomics.com/china/gdp-growth-annual; Mark Magnier, "China's Economic Growth in 2015 Is Slowest in 25 Years," *Wall Street Journal*, January19, 2016, accessed September 14, 2016, http://www.wsj.com/articles/china- economic-growth-slows-to-6-9-on-year-in-2015-1453169398.

98. "China Tops U.S. in Energy Use," *Wall Street Journal*, July 18, 2010, accessed January 2, 2013, http://online.wsj.com/article/SB10001424052748703720504

575376712353150310.html; "China Is Now the World's Largest Net Importer of Petroleum and Other Liquid Fuels—Today in Energy—U.S. Energy Information Administration (EIA)," accessed September 24, 2016, http://www.eia.gov/today-inenergy/detail.cfm?id=15531.

99.    "China—International—Analysis," US Energy Information Administration (EIA), accessed September 24, 2016, https://www.eia.gov/beta/international/analysis.cfm?iso=CHN.

100.    "China Is Now the World's Largest Net Importer of Petroleum and Other Liquid Fuels."

101.    Evan S. Medeiros and M. Taylor Fravel, "China's New Diplomacy," *Foreign Affairs* 82, no. 6 (2003): 22–35.

102.    Steve A. Yetiv and Chunlong Lu, "China, Global Energy, and the Middle East," *Middle East Journal* 61, no. 2 (2007): 199–218.

103.    Robert Bailey, "China and GCC: Growing Ties," *Gulf Business*, April 16, 2013, accessed August 20, 2016, http://gulfbusiness.com/china-and-gcc-growing-ties/.

104.    Golnar Motevalli, "China, Iran Agree to Expand Trade to $600 Billion in a Decade," *Bloomberg.com*, January 23, 2016, accessed September 20, 2016, http://www.bloomberg.com/news/articles/2016-01-23/china-iran-agree-to-expand-trade-to-600-billion-in-a-decade.

105.    Sergei Troush, "China's Changing Oil Strategy and Its Foreign Policy Implications," Brookings Institution, accessed December 10, 2012, http://www.brookings.edu/articles/1999/fall_china_troush.aspx?p=1.

106.    Li Peng was the premier of the PRC between 1987 and 1998, and chairman of the Standing Committee of the National People's Congress, China's top legislative body, from 1998 to 2003.

107.    Troush, "China's Changing Oil Strategy,"

108.    Phillip Andrews-Speed, Stephen Dow, and Zhighuo Gao, "The Ongoing Reforms to China's Government State Sector: The Case of the Energy Industry," *Journal of Contemporary China* 9, no. 23 (2000): 5–20.

109.    Amy Myers-Jaffe and Steven W. Lewis, "Beijing's Oil Diplomacy," *Survival* 44, no. I (2002): 115–34.

110.    Ian Taylor, "China's Oil Diplomacy in Africa," *International Affairs* 82, no. 5 (2006): 937–59.

111.    Erica Downs and Michal Meidan, "Business and Politics in China: The Oil Executive Reshuffle of 2011," *China Security*, no. 19 (2011) accessed December 12, 2012, http://www.chinasecurity.us/index.php?option=com_content&view=article&id=489&Itemid=8.

112.    Erica S. Downs, "The Chinese Energy Security Debate," *China Quarterly* 177 (March 2004): 21–41.

113.    Taylor, "China's Oil Diplomacy in Africa," 937–59.

114.    "CNOOC Limited and its subsidiaries is China's largest producer of offshore crude oil and natural gas and one of the largest independent oil and gas exploration and production companies in the world. The Group mainly engages in exploration, development, production and sales of oil and natural gas. The

Group's core operation areas are Bohai, Western South China Sea, Eastern South China Sea and East China Sea in offshore China. In overseas, the Group has oil and gas assets in Asia, Africa, North America, South America and Oceania." "About us," CNOOC, accessed January, 10, 2013, http://www.cnoocltd.com/enc-noocltd/aboutus/default.shtml.

115. Kevin Sheives, "Beijing's Contemporary Strategy towards Central Asia," *Pacific Affairs* 79, no. 2 (2006): 205–24.

116. Peter S. Goodman, "Big Shift in China's Oil Policy, with Iraq Deal Dissolved by War, Beijing Looks Elsewhere," *Washington Post*, July 13, 2005, accessed November 2012, http://www.washingtonpost.com/wp-dyn/content/article/2005/07/12/AR2005071201546.html.

117. Troush, "China's Changing Oil Strategy."

118. Sheives, "Beijing's Contemporary Strategy towards Central Asia," 205–224.

119. Michael Wines, "Middle East Trip Suggests Change in Policy by China," *New York Times*, January 13, 2012, accessed February 10, 2012, http://www.nytimes.com/2012/01/14/world/asia/wen-jiabao-in-middle-east-as-china-evaluates-oil-policy.html?_r=0.

120. Minxin Pei, "China's Iran Dilemma," *BBC News*, January 19, 2012, accessed April 1, 2012, http://www.bbc.co.uk/news/world-asia-china-16607333.

121. "China Unbowed by US Pressure over Iranian Oil: US Treasury Secretary Fails to Persuade China to Back Sanctions against Iranian Oil Industry during Talks in Beijing," *Aljazeera*, January 11, 2012, accessed January 20, 2012, http://www.aljazeera.com/news/asia-pacific/2012/01/2012110143018616205.html.

122. China imports oil and important minerals from Iran such as iron ore sulfur, copper, zinc, aluminum, granite and marble and exports technology and engineering expertise. "China Ore Stocks Will Sustain Prices Should Iran Face Sanction," Chinamining.org, February 13, 2012, accessed March 9, 2012, http://www.chinamining.org/News/2012-02- 13/1329112697d54284.html.

123. Mahmoud Ghafouri, "China's Policy in the Persian Gulf," Middle East Policy Council, 2013, accessed January 10, 2013, http://www.mepc.org/journal/middle-east-policy-archives/chinas-policy-persian-gulf?print.

124. "MacKay River Commercial Oil Sands Project, Alberta, Canada," *hydrocarbons-technology.com*, 2014, viewed January 3, 2015, http://www.hydrocarbons-technology.com/projects/mackay-river-commercial-oil-sands-project-alberta/.

125. Campbell Clark, "China's Oil-Sands Deal Will Have Lasting Impact," *Globe and Mail*, January 2012, accessed January 3, 2013, http://www.theglobeandmail.com/news/politics/chinas-oil-sands- deal-will-have-lasting-impact/article1357620/.

126. Chris Kahn, "PetroChina passes ExxonMobil in 2011 Crude Oil Production," *USA Today*, March 29, 2012, accessed June 25, 2012, http://www.usatoday.com/money/industries/energy/story/2012-03-29/oil-production-petrochina-exxonmobil/53851876/1.

127. Kahn, "PetroChina Passes ExxonMobil."

128. "Chinese National Oil Companies' Investments: Going Global for Energy," International Energy Agency, November 3, 2014, accessed September 1, 2016, http://www.iea.org/ieaenergy/issue7/chinese-national-oil-companies-investments-going-global-for-energy.html.

129. Stephen Grocer, "Chinese Acquisitions of Foreign Firms Already at Full-Year Record," *Wall Street Journal*, May 10, 2016, accessed September 10, 2016, http://blogs.wsj.com/moneybeat/2016/05/10/chinese-acquisitions-of-foreign-firms-already-at-full-year-record/.

130. Valentina Romei, "China and Africa: Trade Relationship Evolves," *Financial Times*, December 3, 2015, accessed September 28, 2016, https://www.ft.com/content/c53e7f68-9844-11e5-9228-87e603d47bdc.

131. Christopher Alessi and Stephanie Hanson, "Expanding China-Africa Oil Ties," Council on Foreign Relations, February 8, 2012, accessed March 30, 2012, www.cfr.org/china/expanding-china-africa-oil-ties/p9557.

132. Taylor, "China's Oil Diplomacy in Africa," 937–59.

133. Joshua Kurlantzick, *Charm Offensive: How China's Soft Power Is Transforming the World* (Binghamton, NY: Caravan, 2007); Peter Ford, "Chinese Activists Looking to Africa," *Christian Science Monitor*, May 21, 2007, accessed June 19, 2012, http://www.csmonitor.com/2007/0521/p01s02-woap.html.

134. Chris Alden, "Harmony and Discord in China's Africa Strategy: Some Implications for Foreign Policy," *China Quarterly* 199 (2009): 563.

135. Ian Taylor, *China's New Role in Africa* (Boulder, CO: Lynne Rienner, 2010).

## CHAPTER 4

1. Jun Li and Xin Wang, "Energy and Climate Policy in China's Twelfth Five-Year Plan: A Paradigm Shift," *Energy Policy*, no. 42 (2012): 519–28; J. H. L. Voncken, *The Rare Earth Elements: An Introduction* (Delft, NL: Springer, 2015), 110; Seaman, "Rare Earths and Clean Energy."

2. Warren N. Warhol, "Molycorp's Mountain Pass Operations," in *Geology and Mineral Wealth of the California Desert*, ed. D. L. Fife and A.R. Brown (Santa Ana, CA: South Coast Geological Society, 1980), 359–66.

3. "Molycorp's History," Molycorp.com, accessed June 29, 2012, http://www.molycorp.com/about-us/our-history. After Molycorp went bankrupt, the link takes the reader to: http://neomaterials.com/company/our-history/.

4. James B. Hedrick, *Minerals Yearbook Metals and Minerals 1985*, vol. 1 (Washington, DC: US Bureau of Mines), accessed June 10, 2016 http://digital.library.wisc.edu/1711.dl/EcoNatRes.MinYB1985v1.

5. James B. Hedrick, "Rare-Earth Metals," USGS, accessed September 19, 2016, http://minerals.usgs.gov/minerals/pubs/commodity/rare_earths/740798.pdf.

6. C. Wu, Z. Yuan, and G. Bai, "Rare Earth Deposits in China," in *Rare Earth Minerals: Chemistry, Origin and Ore Deposits*, ed. A. P. Jones, F. Wall, and C. T. Williams, The Mineralogical Society Series 7 (London: Chapman and Hall, 1996), 281–310.

7. Hurst, "China's Rare Earth Elements Industry."

8. Pui-Kwan Tse, "China's Rare-Earth Industry" (Open-File Report 2011–1042, USGS, November 12, 2011), http://pubs.usgs.gov/of/2011/1042/.

9. Yasuo Kanazawa and Masaharu Kamitani, "Rare Earth Minerals and Resources in the World," *Proceedings of Rare Earths'04 in Nara, Japan Proceedings of Rare Earths'04* 408–412 (February 9, 2006): 1339–43.

10. J. B. Hendrick, "2005 Minerals Yearbook: Rare Earths," US Geological Survey, http://minerals.usgs.gov/minerals/pubs/commodity/rare_earths/rareemyb05.pdf.

11. Chengyu Wu, "Bayan Obo Controversy: Carbonatites versus Iron Oxide-Cu-Au-(REE-U)," *Resource Geology* 58, no. 4 (2008): 348–54.

12. Pui-Kwan Tse, "China's Rare-Earth Industry."

13. Hurst, "China's Rare Earth Elements Industry."

14. Beijing San Huan New Materials High-Tech Inc. and China National Non-Ferrous Metals Import and Export Corporation used the Sextant Group Inc, to acquire Magnequench. See Joanne Abel Goldman, "The U.S. Rare Earth Industry: Its Growth and Decline," *Journal of Policy History* 26, no. 2 (2014): 139–66, doi:10.1017/S0898030614000013; Charles W. Freeman, "Remember the Magnequench: An Object Lesson in Globalization," *Washington Quarterly* 32, no. 1 (2009): 61–76, doi:10.1080/01636600802545308.

15. Cecilia Jamasmie, "Chinese Rare Earth Alliance Sues Hitachi Metals over Patents," *Mining.com*, August 15, 2014, accessed September 1, 2016, http://www.mining.com/chinese-rare-earth-alliance-sues-hitachi-metals-over-patents-40057/.

16. "Made in China 2025," State Council of the People's Republic of China, English. gov.cn, September 19, 2016, accessed September 19, 2016, http://english.gov.cn/2016special/madeinchina2025/.

17. Don Lee and Elizabeth Douglass, "Chinese Drop Takeover Bid for Unocal," *Los Angeles Times*, August 3, 2005, accessed March 12, 2012, http://articles.latimes.com/2005/aug/03/business/fi-chinaoil3.

18. Steve Lohr, "Unocal Bid Denounced at Hearing," *New York Times*, July 14, 2005, accessed January 5, 2012, http://www.nytimes.com/2005/07/14/business/worldbusiness/14unocal.html.

19. "China Fails in Another Bid for Resources Firm," *The Age*, September 24, 2009, accessed April 12, 2012, http://www.theage.com.au/business/china-fails-in-another-bid-for-resources-firm-20090924-g4l7.html.

20. "China Fails in Another Bid for Resources Firm."

21. Bloomberg News obtained the minutes through an Australia Freedom of Information Act request and published them.

22. Rebecca Keenan, "Australia Blocked Rare Earth Deal on Supply Concerns," *Bloomberg.com*, February 15, 2011, accessed April 20, 2012, http://www.bloomberg.com/news/articles/2011-02-14/australia-blocked-china-rare-earth-takeover-on-concern-of-threat-to-supply.

23. Elisabeth Behrmann, "China's Interest in Arafura Resources' Gains Approval," *The Australian*, May 29, 2009, accessed October 1, 2012, http://www.theaustralian.com.au/business/chinas-interest-in-arafura-resources-gains-approval/story-e6frg8zx-1225717726769.

24. Seaman, "Rare Earths and Clean Energy."

25. Jijo Jacob, "China Tightens Grip on Rare Earths Supply," *International Business Times*, September 27, 2011, accessed May 21, 2012, http://www.ibtimes.com/china-tightens-grip-rare-earths-supply-318698.

26. "Highlights of Proposals for China's 13th Five-Year Plan," Xinhuanet, *English. news.cn*, November 4, 2015, accessed January 3, 2016, http://news.xinhuanet.

com/english/photo/2015-11/04/c_134783513.htm; Wendy Hong, Denise Cheung, and David Sit, "China's 13th Five-Year Plan (2016–2020): Redefining China's Development Paradigm under the New Normal," *Fung Business Intelligence Centre*, November 2015, accessed January 4, 2016, https://www.fbicgroup.com/sites/default/files/China%E2%80%99s%2013th%20Five-Year%20Plan%20(2016-2020)%20Redefining%20China%E2%80%99s%20development%20paradigm%20under%20the%20New%20Normal.pdf.

27.  "Renewables 2017, Global Status Report," Renewable Energy Policy Network for the 21st century (REN21), 2017, accessed June 25, 2017, http://www.ren21.net/wp-content/uploads/2017/06/17-8399_GSR_2017_Full_Report_0621_Opt.pdf.

28.  "China Announces Green Targets," United Press International, accessed May 21, 2012, http://www.upi.com/Business_News/Energy-Industry/2011/03/07/China-announces-green-targets/33531299517620/.

29.  "Renewables 2017, Global Status Report."

30.  Li Junfeng, Cai Fengbo, et al., "2014 China Wind Power Review and Outlook," *Chinese Renewable Energy Industries Association (CREIA), Chinese Wind Energy Association (CWEA), Global Wind Energy Council (GWEC)* accessed May 5, 2015, http://www.gwec.net/wp-content/uploads/2012/06/2014 DDD告 2 英文-20150317.pdf.

31.  "Renewables 2017, Global Status Report"; Pierre Tardieu and Iván Iván Pineda, "Wind in Power: 2016 European Statistics," February 2017, https://windeurope.org/wp-content/uploads/files/about-wind/statistics/WindEurope-Annual-Statistics-2016.pdf.

32.  Jost Wubbeke, "Rare Earth Elements in China: Policies and Narratives of Reinventing an Industry," *Resources Policy* 28 (2013): 384–94.

33.  "Canadian Firms Step Up Search for Rare-Earth Metals," *New York Times*, September 9, 2009, accessed May 21, 2012, http://www.nytimes.com/2009/09/10/business/global/10mineral.html.

34.  "China Cuts Rare Earth Export Quota, May Cause Dispute," *Bloomberg.com*, July 9, 2010, accessed May 15, 2012, http://www.bloomberg.com/news/articles/2010-07-09/china-reduces-rare-earth-export-quota-by-72-in-second-half-lynas-says.

35.  USGS, "Rare Earths," January 2012, accessed May 21, 2012, http://minerals.usgs.gov/minerals/pubs/commodity/rare_earths/mcs-2012-raree.pdf.

36.  USGS, "Rare Earths."

37.  Gareth Hatch, "The First Round of Chinese Rare-Earth Export-Quota Allocations for 2012," *Technology Metals Research*, December 28, 2011, accessed October 1, 2013, http://www.techmetalsresearch.com/2011/12/the-first-round-of-chinese-rare-earth-export-quota-allocations-for-2012/.

38.  Hatch, "First Round."

39.  According to a recent article published in the *China Daily*, the Chinese government is "encouraging" companies across nine critical sectors in the Chinese economy to merge. Rare earths is one of those sectors. Wei Tian, "Ministry Proposes More Mergers in Key Industries," *China Daily*, January 23, 2013, accessed February 17, 2013, http://www.chinadaily.com.cn/m/wudang/2013-01/23/content_16160776.htm.

40. According to the *China Daily*, "In the rare earth industry, Baotou Steel can produce over 7,000 tons of oxide converted hydrometallurgy rare earth product, and
80 varieties and 200 specifications of rare earth products. Baotou Steel can produce
26 varieties and 28 specifications of metallurgical coke and coking by-products."
See, "Baotou Iron and Steel Group (Baotou Steel)," *China Daily*, accessed February
17, 2013, http://www.chinadaily.com.cn/business/2006-11/15/content_734141.htm.

41. Zhang Qi, "Bigger Say Set on Rare Earths Market," *China Daily*, August 10, 2010,
accessed March 2, 2012, http://www.chinadaily.com.cn/china/2010-08/10/con-
tent_11125352.htm.

42. "REFILE-China Approves Rare Earth Industry Consolidation," *Reuters Africa*,
May 14, 2015, accessed June 10, 2015, http://af.reuters.com/article/metalsNews/
idAFL3N0Y55XX20150514.

43. "China Sets 1st Batch of Rare-Earth Ore Mining Quota at 52,500t in 2016," Xinhua
Finance Agency, 5 February 2016, accessed October 17, 2016, http://en.xfafinance.
com/html/Industries/Materials/2016/194395.shtml.

44. "China Six Rare Earth Producers to Produce 99.9% of 1st Batch Production
Quota for 2016," Shanghai Metals Market News, April 8, 2016, accessed October
9, 2016, http://metal.com/newscontent/89553_china-six-rare-earth-producers-to-
produce-99.9-of-1st-batch-production-quota-for-2016: "China Sets 1st Batch."

45. "Lynas Corp Ltd.," *Bloomberg.com*, accessed October 9, 2016, http://www.bloom-
berg.com/quote/LYC:AU.

46. "Arafura Resources Ltd.," *Bloomberg.com*, accessed October 9, 2016, http://www.
bloomberg.com/quote/ARU:AU.

47. "Apple Inc.," *Bloomberg.com*, accessed October 9, 2016, http://www.bloomberg.
com/quote/AAPL:US; "Samsung Electronic Co Ltd.," *Bloomberg.com*, accessed
October 9, 2016, http://www.bloomberg.com/quote/SMSN:LI; "Is Xiaomi Really
Worth $50 Billion?" *Bloomberg.com*, accessed October 9, 2016, http://www.bloom-
berg.com/news/articles/2014-11-04/is-xiaomi-really-worth-50-billion-.

48. According to a news item that appeared in the *China Daily*, smuggling of rare
earths persists. According to Chen Jianxin, deputy director of the administration's anti-smuggling bureau, "The minerals are mainly smuggled to neighboring
countries such as Japan and the Republic of Korea. Chen said the huge demand
from foreign markets and China's high customs duties for rare earths are the main
reasons behind the rise in smuggling. He declined to disclose the latest statistics on the smuggling, but China's first white paper on the rare earths industry,
released by the State Council in June, paints a grim picture. The report said that
in 2011, the amount of rare earths smuggled out of China was 20 percent higher
than the amount of products that legally left the country." Zhang Yan and Wang
Qian, "Smuggling Blights Rare Earths Industry," *China Daily*, December 12, 2012,
accessed February 17, 2013, http://en.people.cn/90778/8051603.html.

49. "China Considers New Invoice System for Rare Earths," *Market Watch*,
November 1, 2011, accessed October 1, 2016, http://www.marketwatch.com/story/
china-considers-new-invoice-system-for-rare-earths-2011-11-01.

50. Mark A. Smith, testimony, "Hearing on Rare Earths Minerals and 21st century Industry" (House Science and Technology Committee Subcommittee

on Investigations and Oversight, Serial No. 111–86, March 16, 2010), accessed March 20, 2012, https://www.gpo.gov/fdsys/pkg/CHRG-111hhrg55844/pdf/ CHRG-111hhrg55844.pdf.

51. "China Said to Add 10,000 Tons to Rare Earths Stockpiles," *Bloomberg.com*, August 5, 2014, accessed January 25, 2015, http://www.bloomberg.com/news/articles/ 2014-08-05/china-said-to-add-10-000-tons-to-rare-earths-stockpiles-1-.

52. "China Said to Add 10,000 Tons to Rare Earths Stockpiles."

53. "Prices of praseodymium-neodymium oxide tumbled 76 percent to 307,500 yuan a ton after reaching a record in mid-2011, according to data from the Shanghai Steelhome Information Technology Co." See "China Said to Add 10,000 Tons to Rare Earths Stockpiles."

54. Mark A. Smith, "China's Rare Earth Export Quotas: Some Observations." Smith's remarks were posted on January 16, 2012, on his *Elementally Clean Blog: Molycorp*, accessed by the author on May 3, 2012, http://www.molycorp. com/News/ElementallyGreenBlog/ElementallyGreenBlogArticle/tabid/795/ ArticleId/196/Chinas-Rare-Earth-Export-Quotas-Some-Observations.aspx. The link was removed following Smith's departure from Molycorp and the company's subsequent bankruptcy.

55. Smith, "China's Rare Earth Export Quotas." As noted in note 55, since Mark Smith left Molycorp and the company's subsequent bankruptcy, this link to the blog in which these comments appeared has been removed.

56. Smith, "China's Rare Earth Export Quotas: Some Observations." As noted in notes 55 and 56, since Mark Smith left Molycorp and the company's subsequent bankruptcy, this link to the blog in which these comments appeared has been removed.

57. Cindy Hurst, "Common Misconceptions of Rare Earth Elements," *Journal of Energy Security*, March 15, 2011, accessed March 12, 2012, http://www.ensec. org/index.php?view=article&catid=114%3Acontent0211&id=290%3Acommon-%20misconceptions-of-rare-earth-%20elements&tmpl=component&print=1&pa ge=&option=com_content&Itemid=374.

58. Lynas Annual General Meeting 2011 Chairman's Address, Lynas Corporation Ltd., November 30, 2011, accessed May 21, 2012, http://www.lynascorp.com/con-tent/upload/files/Presentations/Lynas_2011_AGM_Chairmans_Address_ASX_ FINAL_1053986.pdf.

59. U.S. Department of Energy, "Critical Materials Strategy."

60. U.S. Department of Energy, "Strategic Plan," May 2011, accessed May 21, 2012, http://energy.gov/sites/prod/files/DOE_CMS2011_FINAL_Full.pdf.

61. "Situation and Policies of China's Rare Earth Industry," Information Office of the State Council—The People's Republic of China, June 2010, accessed January 3, 2013, http://english.gov.cn/archive/white_paper/2014/08/23/content_281474983043156. htm.

62. Hurst, "China's Rare Earth Elements Industry."

63. "NEDO: Trilateral EU-Japan-U.S. Conference on Critical Materials," accessed March 26, 2012, http://www.nedo.go.jp/english/event_20120326_index.html.

64. July 2011–via Bloomberg LP, accessed October 2, 2016.

65. February 2012 and December 2012—via Bloomberg LP, accessed October 2, 2016.

66. Abigail Rubenstein, "China to Consolidate Rare Earth Industry through Mergers," *Law360*, accessed May 15, 2015, http://www.law360.com/articles/656429/china-to-consolidate-rare-earth-industry-through-mergers.

67. "China Reforms Resource Tax for Rare Earths," Ministry of Commerce of the People's Republic of China, April 20, 2015, accessed September 19, 2016, http://english.mofcom.gov.cn/article/newsrelease/counselorsoffice/westernasiaandafricareport/201505/20150500959354.shtml.

68. "China Reforms Resource Tax for Rare Earths."

69. "EUROPE 2020: A Strategy for Smart, Sustainable and Inclusive Growth," European Commission, accessed January 10, 2012, http://eunec.vlor.be/detail_bestanden/doc014%20Europe%202020.pdf.

70. "Rare Earth Metal Shortages Could Hamper Deployment of Low-Carbon Energy Technologies" (JRC European Commission news release, November 10, 2011), accessed May 1, 2014.

71. "Report on Critical Raw Materials for the EU," European Commission, May 2014.

72. "SETIS, Strategic Energy Technologies Information System," European Commission, accessed May 10, 2012, http://setis.ec.europa.eu/about-setis/what-is-the-set-plan.

73. "The European Industrial Initiatives are joint large-scale technology development projects between academia, research and industry. The goal of the EIIs is to focus and align the efforts of the Community, Member States and industry in order to achieve common goals and to create a critical mass of activities and actors, thereby strengthening industrial energy research and innovation on technologies for which working at the Community level will add most value. Industrial Initiatives within the SET-Plan include: Wind (The European Wind Initiative), Solar (The Solar Europe Initiative—photovoltaic and concentrated solar power), Electricity Grids (The European Electricity Grid Initiative), Carbon Capture & Storage (The European CO2 Capture, Transport and Storage Initiative), Nuclear Fission (The Sustainable Nuclear Initiative), Bio-energy (The European Industrial Bioenergy Initiative), Smart Cities (Energy Efficiency—The Smart Cities Initiative), Fuel Cells and Hydrogen (Joint Technology Initiative), Nuclear Fusion, ITER (a large-scale scientific experiment that aims to demonstrate that it is possible to produce commercial energy from fusion and F4E (Fusion for Energy)." *SETIS, Industrial Initiatives*, February 18, 2013, accessed February 20, 2013, http://setis.ec.europa.eu/activities/initiatives; "What Is ITER?" *ITER*, accessed September 30, 2016, http://www.iter.org/proj/inafewlines; "Fusion for Energy: Bringing the Power of the Sun to Earth," accessed February 18, 2013, http://fusionforenergy.europa.eu/.

74. "Strategic Energy Technologies," SETIS—European Commission, accessed September 30, 2016, https://setis.ec.europa.eu/technologies.

75. "2030 Climate and Energy Framework: European Commission," accessed September 30, 2016, http://ec.europa.eu/clima/policies/strategies/2030/index_en.htm.

76. Moss, "Potential Risks from Metals Bottlenecks."

77. "EU-US to Collaborate on 'Rare Earths,'" EurActiv.com, July 14, 2010, accessed November 12, 2012, https://www.euractiv.com/section/sustainable-dev/news/

eu-us-to-collaborate-on-rare-earths/; "EU to Step up Raw Materials 'Diplomacy,' " EurActiv.com, June 18, 2010, http://www.euractiv.com/section/sustainable-dev/news/eu-to-step-up-raw-materials-diplomacy/.

78. Abhishek Shah, "Rare Earth 'Geopolitical Problem' to Feature in G-20 Talks Even as USA Looks to Boost Domestic Supply through Legislation," Green World Investor, October 27, 2010, accessed May 10, 2012, http://www.greenworldinvestor.com/2010/10/27/rare-earth-geopolitical-problem-to-feature-in-g-20-talks-even-as-usa-looks-to-boost-domestic-supply-through-legislation/.

79. Keith Bradsher, "Rare Earths Stand Is Asked of G-20," *New York Times*, November 5, 2010, accessed May 10, 2012, http://www.nytimes.com/2010/11/05/business/global/05rarechina.html.

80. "EU Stockpiles Rare Earths as Tensions with China Rise," *Financial Post*, September 6, 2011, accessed April 20, 2012, http://business.financialpost.com/investing/eu-stockpiles-rare-earths-as-tensions-with-china-rise.

81. "Rare Earth Metal Shortages Could Hamper Deployment of Low-Carbon Energy Technologies," Joint Research Center, European Commission's In-House Science Service, accessed February 10, 2012, http://ec.europa.eu/dgs/jrc/index.cfm?id=2300&obj_id=3140&dt_code=PRL&lang=en.

82. The Joint Research Center is the European Commission's in-house science service. Its mission is to provide customer-driven scientific and technical support for the conception, development, implementation, and monitoring of EU policies.

83. "SETIS, Strategic Energy Technologies Information System."

84. "Rare Earth Metal Shortages Could Hamper Deployment of Low-Carbon Energy Technologies."

85. Chikako Mogi and Erik Kirschbaum, "Analysis: Japan, Germany Seek Rare Earth Recycling as Hedge," *Reuters*, November 10, 2010, accessed January 25, 2012, http://www.reuters.com/article/2010/11/10/us-rareearth-recycling-idUSTRE6A90VN20101110.

86. Mogi and Kirschbaum, "Analysis: Japan, Germany Seek Rare Earth Recycling."

87. "Germany, Kazakhstan Sign Strategic Agreement on Rare-Earth Metals," *Jamestown Foundation*, February 14, 2012, accessed September 29, 2016, http://www.jamestown.org/programs/edm/single/?tx_ttnews%5Btt_news%5D=39008&cHash=1f1c8fde7902f4a1852983a887bfb91f.

88. Melissa Eddy, "Germany and Kazakhstan Sign Rare Earths Agreement," *New York Times*, February 8, 2012, accessed March 30, 2012, http://www.nytimes.com/2012/02/09/business/global/germany-and-kazakhstan-sign-rare-earths-agreement.html.

89. "EUROPE 2020: A strategy for Smart, Sustainable and Inclusive Growth."

90. Connie Hedegaard, "Keynote Speech at the Opening of the 'Metropolitan Solutions Conference 2012,' " European Commission," April 23, 2012, accessed May 11, 2012, http://ec.europa.eu/archives/commission_2010-2014/hedegaard/headlines/news/2012-04-23_01_en.htm.

91. "Renewable Energy and Jobs Annual Review, 2016," *IRENA* (International Renewable Energy Agency), accessed September 2, 2016, http://www.irena.org/DocumentDownloads/Publications/IRENA_RE_Jobs_Annual_Review_2016.pdf.

92. "Report on Critical Raw Materials for the EU."

93. For the case of Germany, cf. Maximillian Rech, "Rare Earths and the European Union," in *The Political Economy of Rare Earth Elements: Rising Powers and Technological Change*, International Political Economy Series (Basingstoke, UK: Palgrave Macmillan, 2015), kindle edition.

94. Elaine Kurtenbach, "Japan, US, EU Discuss Rare Earth Supply Security," *Boston. com*, March 28, 2012, accessed November 10, 2012, http://www.boston.com/business/technology/articles/2012/03/28/japan_us_eu_discuss_rare_earth_supply_security/.

95. Yan and Qian, "Smuggling Blights Rare Earths Industry."

96. Marc Humphries, "Rare Earth Elements: The Global Supply Chain," Congressional Research Service, September 6, 2011, accessed May 2, 2012, http://www.fas.org/sgp/crs/natsec/R41347.pdf.

97. Chikako Mogi and Erik Kirschbaum, "Analysis: Japan, Germany Seek Rare Earth Recycling as Hedge," *Reuters*, November 10, 2010, accessed April 27, 2011, http://www.reuters.com/article/2010/11/10/us-rareearth-recycling-idUSTRE6A90VN20101110.

98. "Monitoring Trends in Rare Metals," Japan Oil, Gas and Metals National Corporation (JOGMEC), accessed May 2, 2012, http://www.jogmec.go.jp/english/activities/stockpiling_metal/monitoring.html.

99. Monika Chansoria, "Rare Earth Diplomacy: India and Japan Makes Strategic Partnership to Explore Stakes in Deep-Sea Mining," *Mail Online India*, November 12, 2015, accessed January 5, 2016, http://www.dailymail.co.uk/indiahome/indianews/article-3316143/Rare-earth-diplomacy-India-Japan-makes-strategic-partnership-explore-stakes-deep-sea-mining.html.

100. Sonali Paul, "Japanese Shore Up Cash-Strapped Rare Earths Miner Lynas," *Reuters*, March 13, 2015, accessed September 2, 2015, http://www.reuters.com/article/lynas-corp-debtrenegotiation-idUSL4N0WF2T820150313; Humphries, "Rare Earth Elements."

101. "Sojitz and JOGMEC enter into Definitive Agreements with Lynas Including Availability Agreement to Secure Supply of Rare Earths Products to Japanese Market," JOCMEG, March 30, 2011, accessed March 2, 2014, http://www.jogmec.go.jp/english/news/release/release0069.html.

102. "Trade Statistics," Trade Statistics of Japan Ministry of Finance, accessed October 8, 2016, http://www.customs.go.jp/toukei/srch/indexe.htm.

103. Robert Looney, "Recent Developments on the Rare Earth Front," *World Economics* 12, no. 1 (2011): 47–78.

104. Looney, "Recent Developments on the Rare Earth Front."

105. Jay Solomon, "Clinton Presses Courts Beijing," *Wall Street Journal*, October 29, 2010, accessed December 20, 2011, http://online.wsj.com/article/SB10001424052702303362404575579574019938714.html.

106. "H. Rept. 111-644—RARE EARTHS AND CRITICAL MATERIALS REVITALIZATION ACT OF 2010," webpage, accessed April 5, 2012, https://www.congress.gov/congressional-report/111th-congress/house-report/644/1.

107. "Understanding Rare Earth Metals," GE Lighting, accessed October 10, 2012, http://www.gelighting.com/na/business_lighting/education_resources/rare-earth- elements/downloads/GE_RareEarthFAQ_7.20.11.pdf.

108. "Rare Earth Materials in the Supply Chain," United States Government Accountability Office, GAO, April 1, 2010, accessed September 2, 2012, http://www.gao.gov/assets/100/96654.pdf.

109. In an ironic twist, the video game industry has picked up the threat of the rare earth crisis and has incorporated it in its plot by having the US and China fighting over rare earths. According to an article that ran in *Time*, "Call of Duty: Black Ops II, defined the year 2025 by a rampaging cult, a zombie apocalypse—and a war between the U.S. and China for control of the world's supply of rare earths." See Austin Ramzy, "Precious Holding," *Time*, February 18, 2013, accessed February 18, 2013, http://www.time.com/time/magazine/article/0,9171,2135689,00.html.

110. The steps in the mining process as described in the Government Accountability Office briefing to Congress about rare-earth materials in 2010 are as follows:
     Steps in the Mining Process
     - mining rare earth ore from the mineral deposit;
     - separating the rare earth ore into individual rare earth oxides;
     - refining the rare earth oxides into metals with different purity levels;
     - forming the metals into rare earth alloys; and
     - manufacturing the alloys into components, such as permanent magnets, used in defense and commercial applications.
     "Rare Earth Materials in the Defense Supply Chain," Government Accountability Office, April 1, 2010, accessed May 21, 2012, http://www.gao.gov/assets/100/96654.pdf.

111. "Press Release from Senator Murkowski," Technology & Rare Earth Metals Conference, March 23, 2012, accessed July 1, 2017, http://www.tremcenter.org/index.php?option=com_content&view=article&id=528:press- release-%20from-senator-murkowski&catid=105:latest-news&Itemid=476.

112. US Department of Energy, "Critical Materials Strategy."

113. "Rare Earths Supply-Chain Technology and Resources Transformation Act of 2010 (2010—H.R. 4866)," GovTrack.us, accessed May 20, 2012, https://www.govtrack.us/congress/bills/111/hr4866.

114. "Rare Earths Supply Technology and Resources Transformation Act of 2010 (2010—S. 3521)," GovTrack.us, accessed September 29, 2016, https://www.govtrack.us/congress/bills/111/s3521.

115. Rare Earths and Critical Materials Revitalization Act of 2010 (2010—H.R. 6160).

116. US Department of Energy, "Critical Materials Strategy"; "Energy Critical Elements Renewal Act of 2011 (2011—H.R. 952)," GovTrack.us, accessed May 20, 2012, https://www.govtrack.us/congress/bills/112/hr952.

117. H.R.4883—National Rare-Earth Cooperative Act of 2014, accessed December 2, 2015, https://www.congress.gov/bill/113th-congress/house-bill/4883.

118. H.R. 761 (113th): National Strategic and Critical Minerals Production Act of 2013, https://www.govtrack.us/congress/bills/113/hr761/text.

119.  National Strategic and Critical Minerals Production Act of 2015 (H.R. 1937), accessed September 20, 2016, https://www.govtrack.us/congress/bills/114/hr1937.

120.  Statement of Administration Policy, H.R.1937—National Strategic and Critical Minerals Production Act of 2015, Executive Office of the President, Office of Management and Budget, October 20, 2015, https://www.whitehouse.gov/sites/default/files/omb/legislative/sap/114/saphr1937r_20151020.pdf.

121.  US Department of Energy, "Critical Materials Strategy." "Ames Laboratory to Lead New Research Effort to Address Shortages in Rare Earth and Other Critical Materials," US Department of Energy, January 9, 2013, accessed January 27, 2013, http://energy.gov/articles/ames-laboratory-lead-new-research-effort-address-shortages-rare-earth-and-other-critical.

122.  US Department of Energy, "Critical Materials Strategy."

123.  "Ames Laboratory to Lead New Research Effort to Address Shortages in Rare Earth and Other Critical Materials."

124.  "Annual Report-August 2016," *Critical Materials Institute-US Department of Energy*, August 2016, accessed October 1, 2016, https://cmi.ameslab.gov/sites/default/files/cmi-annual-report-2016.pdf.

125.  "Reconfiguration of the National Defense Stockpile (NDS)," Report to Congress, April 2009, accessed April 1, 2012, http://www.acq.osd.mil/mibp/docs/nds_reconfiguration_report_to_congress.pdf.

126.  "Reconfiguration of the National Defense Stockpile (NDS)."

127.  Shields and Šolar, "Responses to Alternative Forms of Mineral Scarcity," in Dinar, *Beyond Resource Wars*, 254.

128.  "Report of Meeting: Department of Defense Strategic Materials Protection Board," U.S. Department of Defense, December 12, 2008, accessed May 10, 2012, http://www.acq.osd.mil/mibp/docs/report_from_2nd_mtg_of_smpb_12-2008.pdf.

129.  Valerie Bailey Grasso, "Rare Earth Elements in National Defense: Background, Oversight Issues, and Options for Congress," Congressional Research Service, September 5, 2012, accessed January 21, 2013, http://www.fas.org/sgp/crs/natsec/R41744.pdf.

130.  "Press Release," Strategic Materials Advisory Council, March 21, 2013, accessed April 10, 2013,;"2030 Climate & Energy Framework—European Commission," accessed September 30, 2016, http://ec.europa.eu/clima/policies/strategies/2030/index_en.htm.ttp://www.strategicmaterials.org/2013/03/21/strategic-materials-advisory-council-cautions-dod- against-stockpiling-chinese-rare-earths/.

131.  "Mineral Commodity Summaries," USGS, January 2016, accessed May 20, 2016, https://minerals.usgs.gov/minerals/pubs/mcs/2016/mcs2016.pdf. As the book is going to press, it appears that the Defense Logistics Agency has begun to respond to the risk of potential shortages of rare earths. According to the 2017 USGS report The Defense Logistics Agency acquired 8.8 tons of yttrium oxide and planned to add an unspecified quantity of dysprosium metal with a ceiling acquisition of 0.5 tons. "Mineral Commodity Summaries," US Geological Survey, January 2017, accessed June 20, 2017, https://minerals.usgs.gov/minerals/pubs/mcs/2017/mcs2017.pdf.

132. "Rare Earth Materials: Developing a Comprehensive Approach Could Help DOD Better Manage National Security Risks in the Supply Chain" (report to Congressional Committees, United States Government Accountability Office, February 2016), accessed March 1, 2016, http://www.gao.gov/assets/680/675165.pdf.

133. "Rare Earth Materials: Developing a Comprehensive Approach Could Help DOD Better Manage National Security Risks in the Supply Chain."

134. "Executive Order—National Defense Resources Preparedness," Whitehouse. gov, March 16, 2012, https://www.whitehouse.gov/the-press-office/2012/03/16/ executive-order-national-defense-resources-preparedness.

135. "US Imposes Anti-Subsidy Taxes on China's Solar Panels," NDTV, March 3, 2012, accessed May 10, 2012, http://english.ntdtv.com/ntdtv_en/news_ china/2012-03-21/us-imposes-anti-subsidy-taxes-on-china-s-solar-panels. html#video_section; President Barack Obama, Executive Order, National Defense Resources Preparedness, 2012, http://www.whitehouse.gov/the-press-office/2012/03/16/executive-order-national-defense-resources-preparedness; Diane Cardwell, "China's Feud with West on Solar Leads to Tax," *New York Times*, July 18, 2013, accessed March 20, 2014, http://www.nytimes.com/2013/ 07/19/business/energy-environment/chinas-feud-with-west-on-solar-leads-to-tax.html?_r=0.

136. President Obama, "The State of the Union Address," White House.gov, February 12, 2013, accessed February 18, 2013, http://www.whitehouse.gov/photos-and-video/video/2013/02/12/2013- state-union-address-0#transcript.

137. Robert Z. Lawrence, "China and the Multilateral Trading System" (NBER Working Paper No. 12759, National Bureau of Economic Research, Cambridge, MA, December 2006), http://www.nber.org/papers/w12759.

138. Jean-Marc F. Blanchard, "The Dynamics of China's Accession to the WTO: Counting Sense, Coalitions and Constructs," *Asian Journal of Social Science* 41, no. 3–4 (2013): 263–86, doi:10.1163/15685314-12341303.

139. Xiaohui Wu, "No Longer Outside, Not Yet Equal: Rethinking China's Membership in the World Trade Organization," *Chinese Journal of International Law* 10 (2011): 1–49.

140. World Trade Organization, "WTO Successfully Concludes Negotiations on China's Entry" (press release 243, WTO, Geneva, Switzerland, September 17, 2001) accessed December 2, 2014, https://www.wto.org/english/news_e/pres01_ e/pr243_e.htm.

141. Kennan J. Castel-Fodor, "Providing a Release Valve: The U.S.-China Experience with the WTO Dispute Settlement System," *Case Western Reserve Law Review* 64, no. 1 (2013): 201–38.

142. Xiaojun Li, "Understanding China's Behavioral Change in the WTO-Dispute Settlement System: Power, Capacity, and Normative Constraints in Trade Abjudication," *Asian Survey* 52, no. 6 (2012): 1111–137.

143. Kong Qingjiang, "China in the WTO and Beyond: China's Approach to International Institutions," *Tulane Law Review* 88 (2014): 959.

144. Matthew Kennedy, "China's Role in WTO Dispute Settlement," *World Trade Review* 11, no. 4 (2012): 555–89.

145.  Evan S. Medeiros and M. Taylor Fravel, "China's New Diplomacy," *Foreign Affairs* (November/Decembeer, 2003), 22–35; Samuel S. Kim, "International Organizations in Chinese Foreign Policy," *ANNALS of the American Academy of Political and Social Science* 519, no. 1 (January 1, 1992): 140–57; Samuel S. Kim, "China's International Organizational Behaviour," in *Chinese Foreign Policy: Theory and Practice*, ed. Thomas W. Robinson and David Shambaugh (Oxford: Clarendon Press, 1994), 401, 419. Kim argues that China extracts maximum benefits from nuclear disarmament treaties, but has not disarmed itself and thus assumes minimal obligations. See also Pitman B. Potter, "China and the International Legal System: Challenges of Participation," *China Quarterly* 191 (September 2007): 699, 701. Pitman argues that China selectively incorporates international norms most conducive to its economic growth and the preferences of its elites.

146.  Doug Palmer and Sebastian Moffett, "U.S., EU, Japan take on China at WTO over Rare Earths," *Reuters*, March 13, 2012, May 11, 2012, http://www.reuters.com/article/2012/03/13/us-china-trade-eu-idUSBRE82C0JU20120313.

147.  Gareth Hatch, "The WTO Rare Earths Trade Dispute: An Initial Analysis," Technology Metals Research (TMR), March 28, 2012, accessed May 11, 2012, http://www.techmetalsresearch.com/2012/03/the-wto-rare-earths-trade-dispute-an-initial-analysis/.

148.  Palmer and Moffett, "U.S., EU, Japan Take on China at WTO."

149.  "China Responds to WTO Complaint on Rare Earth Mining Cutbacks," *NTD television*, March 21, 2012, accessed May 1, 2012, https://www.youtube.com/watch?v=X8bhCxRXHoc.

150.  Clara Gillispie and Stephanie Pfeiffer, "The Debate over Rare Earths Recent Developments in Industry and the WTO Case" National Bureau of Asian Research, Washington, DC, July 11, 2012) accessed June 10, 2015, http://www.nbr.org/downloads/pdfs/ETA/Hao_Nakanoi_interview_07112012.pdf.

151.  Gillispie and Pfeiffer, "Debate over Rare Earths."

152.  Remarks by the President on Fair Trade, March 13, 2012.

153.  "EU, U.S. Preparing WTO Action against China," *Terra Daily*, June 11, 2009, accessed May 20, 2012, http://www.terradaily.com/reports/EU_US_preparing_WTO_action_against_China_source_999.html; Jill Jusko, "Raw Material Risks, Manufacturers Must Remain Vigilant and Agile in the Face of Volatile Costs and possible Supply Squeezes," *Industry Week*, Oct. 21, 2009, accessed May 20, 2012, http://www.industryweek.com/articles/raw_material_risks_20203.aspx?SectionID=2.

154.  Jennifer M. Freedman, "WTO Rejects Chinese Appeal of Ruling against Mineral Curbs," *Bloomberg Business Week*, January 30, 2012, accessed May 10, 2012, http://www.businessweek.com/news/2012-01-30/wto-rejects-chinese-appeal-of-ruling-against-mineral-curbs.html.

155.  David Stanway, "China Rare Earths Safe from WTO Ruling on Export Curbs," *Reuters*, January 31, 2012, accessed May 20, 2012, http://www.infowars.com/china-rare-earths-safe-from-wto-ruling-on-export-curbs/.

156. Dr. Si Jinsong spoke, on March 14, 2012, at the TREM12 conference, which took place in Washington, DC, on March 13–14, 2012. His presentation followed President Obama's announcement, on March 13, 2012, that the US, the EU, and Japan had requested WTO consultation meetings with China over their rare-earth exports policies.

157. Zhang Yan and Wang Qian, "Smuggling Blights Rare Earths Industry," *People's Daily*, December 12, 2012, accessed February 17, 2013, http://en.people.cn/90778/8051603.html.

158. "Situation and Policies of China's Rare Earth Industry," Information Office of the State Council People's Republic of China, June 2012, accessed December 10, 2012, http://www.miit.gov.cn/n1146285/n1146352/n3054355/n3057569/n3057579/c3590990/part/3590994.pdf.

159. "Situation and Policies of China's Rare Earth Industry."

160. Lu Zhang, et al. "Did China's Rare Earth Export Policies Work? Empirical Evidence from USA and Japan," *Resources Policy*, 43 (2015): 82–90.

161. Zhang, "Did China's Rare Earth Export Policies Work?" 82–90.

162. Ann Lee, *What the U.S. Can Learn from China* (San Francisco: Berrett-Koehler Publishers, 2012), 173.

## CONCLUSION

1. Martin Fackler, "Japan Expands Its Regional Military Role," *New York Times*, November 26, 2012, accessed November 27, 2012, http://www.nytimes.com/2012/11/27/world/asia/japan-expands-its-regional-military-role.html.

2. Will Ripley and Eimi Yamamitsu, "Assertive Japan Poised to Abandon 70 Years of Pacifism," *CNN*, September 18, 2015, accessed September 29, 2016, http://www.cnn.com/2015/09/16/asia/japan-military-constitution/index.html.

3. Nazli Choucri and Robert C. North, "Dynamics of International Conflict: Some Policy Implications of Population, Resources, and Technology," *World Politics* 24 (Spring 1972): 80–83.

4. Valerie Bailey Grasso, "Rare Earth Elements in National Defense: Background, Oversight Issues, and Options for Congress" Congressional Research Service, September 5, 2012, accessed December 6, 2012, http://www.fas.org/sgp/crs/natsec/R41744.pdf.

5. Stephanie Lawson, *International Relations* (Cambridge, UK: Polity, 2012).

# BIBLIOGRAPHY

Abdelal, Rawi, and Adam Segal. "Has Globalization Passed Its Peak?" In *International Politics: Enduring Concepts and Contemporary Issues*, edited by Robert J. Art and Robert Jervis, 340–46. New York: Longman, 2009.

Adshead, Samuel Adrian M. *Salt and Civilization*. New York: St. Martin's, 1991.

Aggarwal, Vinod K., and Sara A. Newland, eds. *Responding to China's Rise*. Vol. 15. Political Economy of the Asia Pacific. Cham, Switzerland: Springer International, 2015.

Alagappa, Muthiah. "Japan's Political and Security Role in the Asia-Pacific Region." *Contemporary Southeast Asia* 10, no. 1 (1998): 17–54.

Alden, Chris. "Harmony and Discord in China's Africa Strategy: Some Implications for Foreign Policy." *China Quarterly* 199 (2009): 563.

Allwood, Julian M., Michael F. Ashby, Timothy G. Gutowski, and Ernst Worrell. "Material Efficiency: Providing Material Services with Less Material Production." *Philisophical Transactions of the Royal Society A 13* 371, no. 1986 (March 2013): 1–15.

Alonso, Elisa, Andrew M. Sherman, Timothy J. Wallington, Mark P. Everson, Frank R. Field, Richard Roth, and Randolph E. Kirchain. "Evaluating Rare Earth Element Availability: A Case with Revolutionary Demand from Clean Technologies." *Environmental Science and Technology* 46, no. 6 (2012): 3406–14.

Andrews-Speed, Philip, Stephen Dow, and Zhighuo Gao. "The Ongoing Reforms to China's Government State Sector: The Case of the Energy Industry." *Journal of Contemporary China* 9, no. 23 (2000): 5–20.

Andrews-Speed, Philip, and Xin Ma. "Energy Production and Social Marginalisation in China." *Journal of Contemporary China* 17, no. 55 (2008): 247–72.

Appert, Nicolas. *L' art de conserver, pendant plusieurs années, toutes les substances animales et végétales*. Paris: Patris et Cie Imprimeurs, 1810.

Auty, Richard M. "The Political Economy of Resource-Driven Growth." *European Economic Review* 45, no. 4–6 (2001): 839–46.

Baldwin, David A. *Econonomic Statecraft*. Princeton, NJ: Princeton University Press, 1958.

Barnett, Harold J., and Chandler Morse. *Scarcity and Growth*. Baltimore: Johns Hopkins Press for Resources for the Future, 1963.

Baumol, William J., and Alan S. Blinder. *Economics: Principles and Policy*. Mason, OH: South-Western Cengage Learning, 2011.

Bergier, Jean-François. *Une Histoire du sel*. Fribourg: Presses Universitaires de France, 1982.

Betts, Richard K., and Thomas J. Christensen. "China: Getting the Questions Right." *National Interest*, no. 62 (2000): 17–29.

Binnemans, Koen, Peter Tom Jones, Bart Blanpain, Tom Van Gerven, Yongxiang Yang, Allan Walton, and Matthias Buchert. "Recycling of Rare Earths: A Critical Review." *Journal of Cleaner Production* 51 (July 2013): 1–22. doi:10.1016/j.jclepro.2012.12.037.

Blanchard, Jean-Marc F. "The Dynamics of China's Accession to the WTO: Counting Sense, Coalitions and Constructs." *Asian Journal of Social Science* 41, no. 3 (2013): 263–86.

Boserup, Ester. *The Conditions of Agricultural Growth: The Economics of Agrarian Change under Population Pressure*. London: Allen & Unwin, 1965.

Brzeszinski, Zbigniew. *The Grand Chessboard*. New York: Basic, 1997.

Cáceres, Sigfrido Burgos, and Sophal Ear. "The Geopolitics of China's Global Resources Quest." *Geopolitics* 17, no. 1 (2012): 47–79.

Cao, Wensheng, and Christoph Bluth. "Challenges and Countermeasures of China's Energy Security." *Energy Policy*, no. 53 (2013): 381–88.

Cardarelli, Francois. *Materials Handbook: A Concise Desktop Reference*. London: Springer, 2008.

Casarini, Nicola. *Remaking Global Order: The Evolution of Europe-China Relations and Its Implications for East Asia and the United States*. New York: Oxford University Press, 2009.

Castel-Fodor, Kennan J. "Providing a Release Valve: The US-China Experience with the WTO Dispute Settlement System." *Case Western Reserve Law Review* 64 (2013): 201.

Chadha, Yogesh. *Gandhi: A Life*. New York: John Wiley, 1997.

Chai, Winberg. "The Ideological Paradigm Shifts of China's World Views: From Marxism-Leninism-Maoism to the Pragmatism-Multilateralism of the Deng-Jiang-Hu Era." *Asian Affairs* 30, no. 3 (2003): 163–75.

Chen, Dingding, and Xiaoyu, Pu. "Debating China's Assertiveness." *International Security* 38, no. 3 (2013): 176–83.

Cheng, Joseph Y. S. "A Chinese View of China's Energy Security." *Journal of Contemporary China* 17, no. 55 (2008): 297–317.

Chiang, Tao-Chang. "The Salt Trade in Ch'ing China." *Modern Asian Studies* 17, no. 2 (1983): 218.

Choucri, Nazli, and Robert C. North. "Dynamics of International Conflict: Some Policy Implications of Population, Resources, and Technology." In "Theory and Policy of International Relations." Supplement. *World Politics* 24 (1972): 80–122.

Christensen, Thomas J. *The China Challenge: Shaping the Choices of a Rising Power*. New York: W. W. Norton, 2016.

Chung, Chien-Peng. "Japan's Involvement in Asia-Centered Regional Forums in the Context of Relations with China and the United States." *Asian Survey* 52, no. 3 (2011): 407–448.

Click, Reid, and Robert J. Weiner. "Resource Nationalism Meets the Market: Political Risk and the Value of Petroleum Reserves." *Journal of International Business Studies* 41, no. 5 (2010): 783.

Collier, Paul, and Anke Hoeffler. "Greed and Grievance in Civil War." *Oxford Economic Papers* 56, no. 4 (2004): 439–58.

Constantin, Christian. "Understanding China's Energy Security." *World Political Science Review* 3 (2007): 1–15.

Deng, Yong. *China Rising: Power and Motivation in Chinese Foreign Policy*. Lanham, MD: Rowman and Littlefield, 2005.

———. *China's Struggle for Status the Realignment of International Relations*. Cambridge: Cambridge University Press, 2008.

Devadason, Evelyn S. "The Trans-Pacific Partnership (TPP): The Chinese Perspective." *Journal of Contemporary China* 23, no. 87 (2014): 462–79. doi:10.1080/10670564. 2013.843890.

Downing, Theodore E. *Avoiding New Poverty: Mining-Induced Displacement and Resettlement*. Report. No. 58. International Institute for Environment and Development London, 2002.

Downs, Erica S. "The Chinese Energy Security Debate." *China Quarterly* 177 (2004): 21–41.

Downs, Erica, and Michal Meidan. "Business and Politics in China: The Oil Executive Reshuffle of 2011." *China Security* 19 (2011): 3–21.

Dresner, Simon. *The Principles of Sustainability*. London: Earthscan Publications, 2008.

Drezner, Daniel W. *The Sanction Paradox: Economic Statecraft and International Relations*. Cambridge: Cambridge University Press, 1999.

———. "China's Africa Strategy." *Current History* 105, no. 691 (2006): 219.

Eisenman, Joshua, Eric Heginbotham, and Derek Mitchell. *China and the Developing World: Beijing's Strategy for the Twenty-First Century*. Armonk, NY: M. E. Sharpe, 2007.

Enuka, Chuka. "The Forum on China-Africa Cooperation (Focac): A Framework for China's Re-engagement with Africa in the 21st century." *E-BANGI: Journal of Social Sciences and Humanities* 6, no. 2 (2011): 220.

Eskew, Garnett. *Salt, The Fifth Element: The Story of a Basic American Industry*. Chicago: J. G. Ferguson, 1948.

Flad, Rowan K. *Salt Production and Social Hierarchy in Ancient China: An Archaeological Investigation of Specialization in China's Three Gorges*. Cambridge: Cambridge University Press, 2011.

Flournoy, Michèle, and Janine Davidson. "Obama's New Global Posture: The Logic of U.S. Foreign Deployments." *Foreign Affairs* 91, no. 4 (2012): 54–63.

Foot, Rosemary. *China, the United States, and Global Order*. New York: Cambridge University Press, 2010.

Freeman, Charles W. "Remember the Magnequench: An Object Lesson in Globalization." *Washington Quarterly* 32, no. 1 (2009): 61–76. doi:10.1080/01636600802545308.

Friedberg, Aaron L. *A Contest for Supremacy: China, America, and the Struggle for Mastery in Asia*. New York: W. W. Norton, 2011.

Frosch, Robert A., and Nicholas E. Gallopoulos. "Strategies for Manufacturing." *Scientific American* 261, no. 3 (1989): 144–52.

Gale, Esson M. "Public Administration of Salt in China: A Historical Survey." *Annals of the American Academy of Political and Social Science* 152 (1930): 241–51.

Gautier, Catherine. *Oil, Water, and Climate: An Introduction.* New York: Cambridge University Press, 2008.

Gilbar, Gad. *The Middle East Oil Decade and Beyond: Essays in Political Economy.* London: Frank Cass, 1997.

Gilmore, Harlan W. "Cultural Diffusion via Salt." *American Anthropologist* 57, no. 5 (1955): 1011–15.

Goldman, Joanne Abel. "The U.S. Rare Earth Industry: Its Growth and Decline." *Journal of Policy History* 26, no. 2 (2014): 139–66. doi:10.1017/S0898030614000013.

Golev, Artem, Margaretha Scott, Peter D. Erskine, Saleem H. Ali, and Grant R. Ballantyne. "Rare Earths Supply Chains: Current Status, Constraints and Opportunities." *Resources Policy* 41 (2014): 52–59.

Goodstein, David. *Out of Gas: The End of the Age of Oil.* New York: W. W. Norton, 2005.

Gordon, B. R., M. Bertram, and E. T. Graedel. "On the Sustainability of Metal Supplies: A Responce to Tilton and Lagos." *Resources Policy* 32 (2007): 24.

Graedel, T. E., and Braden R. Allenby. *Industrial Ecology and Sustainable Engineering.* Englewood Cliffs: Prentice Hall, 1995.

Graedel, T. E., and R. J. Lifset. "Industrial Ecology's First Decade." In *Taking Stock of Industrial Ecology,* edited by Roland Clift and Angela Druckman, 3–20. New York: Springer International Publishing, 2016. doi:10.1007/978-3-319-20571-7_1.

Greenfield, Aaron, and T. E. Graedel. "The Omnivorous Diet of Modern Technology." *Resources, Conservation and Recycling* 74 (May 2013): 1–7. doi:10.1016/j.resconrec.2013.02.010.

Gschneidner, K. A., L. Eyring, and M. B. Maple. *Handbook on the Physics and Chemistry of Rare Earths: High Temperature Rare Earths Superconductors—I.* North Holland: Elsevier, 2000.

Gupta, C. K., and N. Krishnamurthy. *Extractive Metallurgy of Rare Earths.* Boca Raton, FL: CRC Press, 2005.

Hachigian, Nina, ed. *Debating China: The U.S.-China Relationship in Ten Conversations.* Oxford: Oxford University Press, 2014.

Haglund, David G. "The New Geopolitics of Minerals: An Inquiry into the Changing International Significance of Strategic Minerals." *Political Geography Quarterly,* no. 5 (1986): 227.

Hagström, Linus. "'Power Shift' in East Asia? A Critical Reappraisal of Narratives on the Diaoyu/Senkaku Islands Incident in 2010." *Chinese Journal of International Politics* 5, no. 3 (2012): 267–97. doi:10.1093/cjip/pos011.

Har-El, Menashe. "The Routes of Salt, Sugar and Balsam Caravans in the Judean Dessert." *GeoJournal* 2, no. 6 (1978): 549–556.

Hendrick, James B. "Rare-Earth Metal Prices in the USA Ca. 1960–1994." *Journal of Alloys and Compounds,* 250 (1997): 61.1–61.7.

Hill, Steven. *Europe's Promise: Why the European Way Is the Best Hope in an Insecure Age.* Berkeley: University of California Press, 2010.

Hirth, Friedrich. *The Ancient History of China.* New York: Columbia University Press, 1908.

Hocquet, Jean-Claude. "Capitalisme marchand et classe marchande à Venise au temps de la Renaissance." *Annales Histoire, Sciences Sociales,* no. 2 (1979): 279–304.

Homer-Dixon, Thomas. "Cornucopians and Neo-Malthusians." In *International Politics, Enduring Concepts and Contemporary Issues*, edited by J. Art Robert and Robert Jervis, 522–24. New York: Pearson/Longman, 2009.

——. *Environment, Scarcity, and Violence*. Princeton, NJ: Princeton University Press, 1999.

Huan, K 'uan. *Discourses on Salt and Iron: A Debate on State Control of Commerce and Industry in Ancient China, Chapters I–XIX*. Leiden, NL: E. J. Brill, 1931.

Huliaras, Asteris. "The Illusion of Unitary Players and the Fallacy of Geopolitical Rivalry: The European Union and China in Africa." *The Round Table* 101, no. 5 (2012): 425–34.

Humphreys, Macartan, Jeffrey D. Sachs, and Joseph E. Stieglitz, eds. *Escaping the Resource Curse*. New York: Columbia University Press, 2007.

Jiemian, Yang. "The Change of America's Power and Re-structure of International System." *International Studies*, no. 2 (2012): 57.

Johnston, Alastair Iain. "How New and Assertive Is China's New Assertiveness?" *International Security* 37, no. 4 (2013): 7–48.

Johnstone, Christopher B. "Paradigms Lost: Japan's Asia Policy in a Time of Growing Chinese Power." *Contemporary Southeast Asia* 21, no. 3 (1999): 365–85.

Jones, A. P., F. Wall, and C. T. Williams. *Rare Earth Minerals: Chemistry, Origin and Ore Deposits*. The Mineralogical Society Series 7. London: Chapman and Hall, 1996.

Juul, Kristine. "Transhumance, Tubes, and Telephones: Drought Related Migration as a Process of Innovation." In *Beyond Territory and Scarcity: Exploring Conflicts over Natural Resource Scarcities*, edited by Quentin Gausset, Michael Whyte, and Michael Birch-Thomsen, 112–34. Uppsala, Sweden: Nordic Africa Institute, 2005.

Kahl, Colin. *States, Scarcity, and Civil Strife in the Developing World*. Princeton, NJ: Princeton University Press, 2008.

Kalantzakos, Sophia. *EU, US and China Tackling Climate Change: Policies and Alliances for the Anthropocene*. Abingdon, UK: Routledge, 2017.

Kanazawa, Yasuo, and Masaharu Kamitani. "Rare Earth Minerals and Resources in the World." *Proceedings of Rare Earths'04 in Nara, Japan Proceedings of Rare Earths'04* 408–412 (2006): 1339–43. doi:10.1016/j.jallcom.2005.04.033.

Kennedy, Matthew. "China's Role in WTO Dispute Settlement." *World Trade Review* 11, no. 4 (2012): 555–89. doi:10.1017/S1474745612000365.

Kennedy, Paul. *The Rise and Fall of the Great Powers: Economic Change and Military Conflict from 1500 to 2000*. New York: Vintage, 1987.

Keohane, Robert O. *After Hegemony: Cooperation and Discord in the World Political Economy*. Princeton, NJ: Princeton University Press, 1984.

——. "International Institutions: Can Interdependence Work?" In *International Politics, Enduring Concepts and Contemporary Issues*, edited by Robert J. Art and Robert Jervis, 150–58. New York: Pearson/Longman, 2010.

Kesler, S. E. *Mineral Resources, Economics and the Environment*. New York: Macmillan, 1996.

Khawlie, M. R. *Beyond the Oil Era? Arab Mineral Resources and Future Development*. London New York: Mansell, 1990.

Kim, Samuel S. "China's International Organizational Behaviour." In *Chinese Foreign Policy: Theory and Practice*, edited by Thomas W. Robinson and David Shambaugh, 401–34. Oxford: Clarendon Press, 1994.

———. "International Organizations in Chinese Foreign Policy." *Annals of the American Academy of Political and Social Science* 519, no. 1 (1992): 140–57.

Klare, Michael. *Resource Wars: The New Landscape of Global Conflict*. New York: Metropolitan, 2002.

Korin, Anne, and Gal Luft. *Turning Oil into Salt: Energy Independence through Fuel Choice*. Charleston, SC: BookSurge, 2009.

Kostoglou, M., A. G. Konstandopoulos, and H. Burtscher. "Size Distribution Dynamics of Fuel-Borne Catalytic Ceria Nanoparticle." *Journal of Aerosol Science* 38, no. 6 (2007): 604–11.

Kurlansky, Mark. *Birdseye: The Adventures of a Curious Man*. New York: Anchor, 2012.

———. *Salt: A World History*. New York: Penguin, 2003.

Kurlantzick, Joshua. *Charm Offensive: How China's Soft Power Is Transforming the World*. Binghamton, NY: Caravan, 2007.

Lawson, Stephanie. *International Relations*. Cambridge, UK: Polity, 2012.

Le Billon, Philippe. ed. *The Geopolitical Economy of Resource Wars*. Geopolitics of Resource Wars. London: Frank Cass, 2005.

Lee, Ann. *What the U.S. Can Learn from China: An Open-Minded Guide to Treating Our Greatest Competitor as Our Greatest Teacher*. San Francisco: Berrett-Koehler, 2012.

Li, Jun, and Xin Wang. "Energy and Climate Policy in China's Twelfth Five-Year Plan: A Paradigm Shift." *Energy Policy*, no. 42 (2012): 519–28.

Li, Xiaojun. "Understanding China's Behavioral Change in the WTO Dispute Settlement System." *Asian Survey* 52, no. 6 (2012): 1111–37. doi:10.1525/as.2012.52.6.1111.

Lieber, Robert. *The Oil Decade: Conflict and Cooperation in the West*. New York: Praeger, 1983.

Lipschutz, Ronnie. *When Nations Clash: Raw Materials, Ideology, and Foreign Policy*. New York: Ballinger, 1989.

Loewe, Michael, and Edward L. Shaughnessy. *The Cambridge History of Ancient China: From the Origins of Civilization to 221 BC*. New York: Cambridge University Press, 1999.

Lomborg, Bjørn. "Resource Constraints or Abundance?" In *Environmental Conflict*, edited by Paul F. Diehl and Nils P. Gleditsch, 125–52. Boulder, CO: Westview, 2000.

Lonn, Ella. *Salt as a Factor in the Confederacy*. University: University of Alabama Press, 1965.

Looney, Robert. "Recent Developments on the Rare Earth Front." *World Economics* 12, no. 1 (2011): 47–78.

Lorentzou, S., C. Pagkoura, A. Zygogianni, G. Kastrinaki, and A. G. Kostandopoulos. "Catalytic Nano-Structured Materials for Next Generation Diesel Particulate Filters." *SAE International Journal of Materials and Manufacturing* 1, no. 1 (2009): 181–98.

Lucas, Jacques, Pierre Lucas, Thierry Le Mercier, Alain Rollat, and William G. Davenport. *Rare Earths: Science, Technology, Production and Use*. 1st edition. Amsterdam: Elsevier, 2014.

Luo, Yang. "The Inexorable Rise of China's NdFeB Magnet Industry." *Metal Powder Report* 63, no. 11 (2008): 8–10.

MacFarquhar, Roderick. *The Politics of China: Sixty Years of the People's Republic of China.* Cambridge: Cambridge University Press, 2011.

Malthus, Thomas. *An Essay on the Principle of Population.* London: J. Johnson, 1798.

Mandel, Robert. *Conflict over the World's Resources: Background, Trends, Case Studies, and Considerations for the Future.* New York: Greenwood, 1988.

Massari, Stefania, and Marcello Ruberti. "Rare Earth Elements as Critical Raw Materials: Focus on International Markets and Future Strategies." *Resources Policy* 38, no. 1 (2013): 36–43.

Maugeri, Leonardo. *The Age of Oil: The Mythology, History, and Future of the World's Most Controversial Resource.* Westport, CT: Praeger, 2006.

Medeiros, Evan S., and M. Taylor Fravel. "China's New Diplomacy." *Foreign Affairs* 82, no. 6 (2003): 22–35.

Michishita, Narushige, and Richard J. Samuels. "Hugging and Hedging." In *Worldviews of Aspiring Powers*, edited by Henry R. Nau and Deepa Ollapally, 146–76. Oxford University Press, 2012.

Mogi, Chikako, and Erik Kirschbaum. "Analysis: Japan, Germany Seek Rare Earth Recycling as Hedge." *Reuters*, November 10, 2010. http://www.reuters.com/article/2010/11/10/us-rareearth-recycling-idUSTRE6A90VN20101110.

Mortimer, Charles E. *Chemie: das Basiswissen der Chemie; 123 Tabellen.* Stuttgart: Thieme, 2001.

Mortimore, Michael. "Social Resilience in African Dryland Livelihoods: Deriving Lessons for Policy." In *Beyond Territory and Scarcity: Exploring Conflicts over Natural Resource Scarcities*, edited by Quentin Gausset, Michael Whyte, and Michael Birch-Thomsen, 46–69. Uppsala, Sweden: Nordic Africa Institute, 2005.

Moss, R. L., E. Tzimas, H. Kara, P. Willis, and J. Kooroshy. "The Potential Risks from Metals Bottlenecks to the Deployment of Strategic Energy Technologies." *Energy Policy*, 55 (2013): 556–64.

Multhauf, Robert P. *Neptune's Gift: A History of Common Salt.* Baltimore: Johns Hopkins University Press, 1978.

Myers-Jaffe, Amy, and Stephen W. Lewis. "Beijin's Oil Diplomacy." *Survival* 44, no. 1 (2002): 115–34.

Neelameggham, Neale R., Shafiq Alam, Harald Oosterhof, Animesh A. Jha, and Shijie Wang. *Rare Metal Technology 2014.* Hoboken, New Jersey: John Wiley & Sons, 2014.

Nickels, Liz. "The Growing Pull of Rare Earth Magnets." *Metal Powder Report* 65, no. 2 (2010): 6–8.

Nugent, Neill. *The Government and Politics of the European Union.* Basingstoke, UK: Palgrave Macmillan, 2010.

Nye, Joseph. *Soft Power: The Means to Success in World Politics.* New York: Public Affairs, 2004.

Odell, Peter R. "Towards a Geographically Reshaped World Oil Industry." *World Today* 37, no. 12 (1981): 447–53.

Parker, E. H. "The Chinese Salt Trade, an Opening for British Enterprise." *Economic Journal* 10 (1900): 116–25.

Potter, Pitman B. "China and the International Legal System: Challenges of Participation." *China Quarterly* 191 (September 2007). doi:10.1017/S0305741007001671.

Princen, Thomas. *The Logic of Sufficiency*. Cambridge, MA: MIT Press, 2005.

Qingjiang, Kong. "China in the WTO and Beyond: China's Approach to International Institutions." *Tulane Law Review* 88 (2014): 959.

Rajan, Raghuram. "The Great Game Again?" *Finance and Development* 43, no. 4 (2006): 54–55.

Rech, Maximilian. "Rare Earth Elements and the European Union." In *The Political Economy of Rare Earth Elements*, edited by Ryan David Kiggins, 62–84. International Political Economy Series. Basingstoke, UK: Palgrave Macmillan, 2015. doi:10.1057/9781137364241_4.

Reuter, Markus, and Antoinette van Schaik. "Transforming the Recovery and Recycling of Nonrenewable Resources." In *Linkages of Sustainability*, edited by Thomas Graedel and Ester van der Voet, 150–62. Cambridge, MA: MIT Press, 2010.

Roach, Stephen Samuel. *Unbalanced: The Codependency of America and China*. New Haven, CT: Yale University Press, 2014.

Ross, Michael L. "Blood Barrels: Why Oil Wealth Fuels Conflict." *Foreign Affairs* 87, no. 3 (2008): 2–8.

Sachs, Jeffrey D., and Andrew M. Warner. "The Curse of Natural Resources." *European Economic Review* 45, no. 4–6 (2001): 827–38.

Sadeghbeigi, Reza. *Fluid Catalytic Cracking Handbook: An Expert Guide to the Practical Operation, Design, and Optimization of FCC Units*. Amsterdam: Elsevier, 2012.

Segal, Adam "Chinese Economic Statecraft and the Political Economy of Asian Security." In *China's Rise and the Balance of Influence in Asia*, edited by William Keller and Thomas Rawski, 146–61. Pittsburgh: University of Pittsburgh Press, 2007.

Sheives, Kevin. "Beijing's Contemporary Strategy Towards Central Asia." *Pacific Affairs* 79, no. 2 (2006): 205–24.

Shields, Deborah J., and Slavko V. Šolar. "Responses to Alternative Forms of Mineral Scarcity: Conflict and Cooperation." In *Beyond Resource Wars: Scarcity, Environmental Degradation, and International Cooperation*, edited by Shlomi Dinar, 239–85. Cambridge, MA: MIT Press, 2011.

Simon, Julian L. *The State of Humanity*. Cambridge, MA: Blackwell, 1995.

———. *The Ultimate Resource 2*. Princeton, NJ: Princeton University Press, 1996.

Sohn, Yul. "Japan's New Regionalism: China Shock, Values, and the East Asian Community." *Asian Survey* 50, no. 3 (2010): 497–519.

Song, Guoyou, and Wen Jin Yuan. "China's Free Trade Agreement Strategies." *Washington Quarterly* 35, no. 4 (2012): 107–19.

Spence, Jonathan. *The Gate of Heavenly Peace: The Chinese and Their Revolution, 1895–1980*. New York: Penguin, 1981.

Spencer, Joseph Earle. "Salt in China." *Geographical Review* 25, no. 3 (1935): 353–66. doi:10.2307/209305.

Stern, Roger. "Oil Market Power and United States National Security." *PNAS* 103, no. 5 (2006): 1650–55.

Taylor, Ian. *China's New Role in Africa*. Boulder, CO: Lynne Rienner, 2010.

———. "China's Oil Diplomacy in Africa." *International Affairs* 82, no. 5 (2006): 1468–2346.

Tilton, John E. "Exhaustible Resources and Sustainable Development: Two Different Paradigms." *Resources Policy* 22, no. 1–2 (1996): 91–97.

———. *On Borrowed Time? Assessing the Threat of Mineral Depletion*. Washington DC: RFF Press, 2003.

Voncken, J. H. L. *The Rare Earth Elements: An Introduction*. Delft, NL: Springer, 2015.

Wang, Chi. *Obama's Challenge to China: The Pivot to Asia*. Farnham, Surrey, UK: Ashgate, 2015.

Wang, Dianzuo. "Perspectives on China's Mining and Mineral Industry." In *A Review on Indicators of Sustainability for the Mineral Extraction Industries*, edited by Roberto C. Villas-Boas, Deborah Shields, Slavko V. Šolar, Paul Anciaux, and Güven Önal. Rio de Janeiro: CETEM/MCT/ CNPq/CYTED/IMPC, 2005.

Wang, Vincent Wei-cheng. "China's Economic Statecraft toward Southeast Asia: Free Trade Agreement and 'Peaceful Rise.'" *American Journal of Chinese Studies* 1, no. 1 (April 2006): 5–34.

Warhol, Warren N. "Molycorp's Mountain Pass Operations." In *Geology and Mineral Wealth of the California Desert*, edited by D. L. Fife and A. R. Brown, 359–66. Santa Anna, CA: South Coast Geological Society, 1980.

Wu, C., Z. Yuan, and G. Bai. "Rare Earth Deposits in China." In *Rare Earth Minerals: Chemistry, Origin and Ore Deposits*, edited by A. P. Jones, F. Wall, and C. T. Williams, 281–310. The Mineralogical Society Series 7. London: Chapman and Hall, 1996.

Wu, Chengyu. "Bayan Obo Controversy: Carbonatites versus Iron Oxide-Cu-Au-(REE-U)." *Resource Geology* 58, no. 4 (2008): 348–54. doi:10.1111/j.1751-3928.2008.00069.x.

Wu, Xiaohui. "No Longer Outside, Not Yet Equal: Rethinking China's Membership in the World Trade Organization." *Chinese Journal of International Law* 10 (2011): 1–49. Available at SSRN: https://ssrn.com/abstract=1743559.

Wübbeke, Jost. "Rare Earth Elements in China: Policies and Narratives of Reinventing an Industry." *Resources Policy* 38, no. 3 (2013): 384–94.

Xiangyang, Li, Zhongguo Jueqi Guocheng, and Zhong De Zhongda Tiaozhan. "TPP: A Serious Challenge for China's Rise." *International Economic Review* 2 (2012): 17–27.

Xiaochuan, Zhou. "Reform the International Monetary System." *BIS Review* 41 (2009): 1–3.

Xu, Guochang, Junya Yano, and Shin-ichi Sakai. "Scenario Analysis for Recovery of Rare Earth Elements from End-of-Life Vehicles." *Journal of Material Cycles and Waste Management* 18, no. 3 (2016): 469–82. doi:10.1007/s10163-016-0487-y.

Yetiv, Steve A., and Chunlong Lu. "China, Global Energy, and the Middle East." *Middle East Journal* 61, no. 2 (2007): 199–218.

Young, J. E. "Mining the Earth." In *State of the World*, edited by Lester R. Brown, Edward C. Wolf, and Linda Stark, 110–18. New York: W. W. Norton, 1992.

Zepf, Volker. *Rare Earth Elements*. Springer Theses. Heidelberg: Springer, 2013. http://link.springer.com/10.1007/978-3-642-35458-8.

Zhang, Lu, Qing GUO, Junbiao Zhang, Yong Huang, and Tao Xiong. "Did China's Rare Earth Export Policies Work? Empirical Evidence from USA and Japan." *Resources Policy* 43, no. C (2015): 82–90.

Zweig, David, and Jianhai, Bi. "China's Global Hunt for Energy." *Foreign Affairs* 84, no. 5 (2005): 25–38.

Page references for figures are indicated by *f* and for tables by *t*.

Printed in the USA/Agawam, MA
October 6, 2020

762310.102